SAMARIUM

CHEMICAL PROPERTIES, OCCURRENCE AND POTENTIAL APPLICATIONS

CHEMISTRY RESEARCH AND APPLICATIONS

Additional books in this series can be found on Nova's website under the Series tab.

Additional e-books in this series can be found on Nova's website under the e-book tab.

CHEMISTRY RESEARCH AND APPLICATIONS

SAMARIUM

CHEMICAL PROPERTIES, OCCURRENCE AND POTENTIAL APPLICATIONS

KAITLYN R. DANFORD
EDITOR

New York

For permission to use material from this book please contact us:
Telephone 631-231-7269; Fax 631-231-8175
Web Site: http://www.novapublishers.com

NOTICE TO THE READER

The Publisher has taken reasonable care in the preparation of this book, but makes no expressed or implied warranty of any kind and assumes no responsibility for any errors or omissions. No liability is assumed for incidental or consequential damages in connection with or arising out of information contained in this book. The Publisher shall not be liable for any special, consequential, or exemplary damages resulting, in whole or in part, from the readers' use of, or reliance upon, this material. Any parts of this book based on government reports are so indicated and copyright is claimed for those parts to the extent applicable to compilations of such works.

Independent verification should be sought for any data, advice or recommendations contained in this book. In addition, no responsibility is assumed by the publisher for any injury and/or damage to persons or property arising from any methods, products, instructions, ideas or otherwise contained in this publication.

This publication is designed to provide accurate and authoritative information with regard to the subject matter covered herein. It is sold with the clear understanding that the Publisher is not engaged in rendering legal or any other professional services. If legal or any other expert assistance is required, the services of a competent person should be sought. FROM A DECLARATION OF PARTICIPANTS JOINTLY ADOPTED BY A COMMITTEE OF THE AMERICAN BAR ASSOCIATION AND A COMMITTEE OF PUBLISHERS.

Additional color graphics may be available in the e-book version of this book.

Library of Congress Cataloging-in-Publication Data

Samarium : chemical properties, occurrence and potential applications / [edited by] Kaitlyn R. Danford.
   pages cm. -- (Chemistry research and applications)
 Includes bibliographical references and index.
 ISBN 978-1-63321-045-5 (softcover)
 1. Samarium. I. Danford, Kaitlyn R., editor.
 QD181.S65S26 2014
 546'.415--dc23
                    2014018029

Published by Nova Science Publishers, Inc. † New York

CONTENTS

PREFACE

Samarium is a lanthanide that preferentially exists in the +3 oxidation state. Samarium compounds are useful as catalysts which have been employed in a large number of chemical reactions. Catalysts are very important in chemistry and are used in more than 80% of industrial processes. Since chemical processes transform raw materials into products, it is important that these reactions are relatively quick and selective so they can be considered for an industrial scale. This book discusses samarium compounds, and continues to provide information on the Sm-complex/polymer composite; activation procedures of SmI_2 and its utilization for organic reactions; a comparison of Sm(III) and Cr(VI) ions; and potential applications of samarium as a dopant element.

Chapter 1 – In catalysis, a relatively small amount of a material (named catalyst) is added to increase the rate of the reaction, without it being consumed, thus allowing the speed up of a reaction and the selective formation of a product. Catalysts are very important in chemistry and are used in more than 80 % of industrial processes.

Samarium is a lanthanide that preferentially exists in the +3 oxidation state. Samarium compounds are useful as catalysts which have been employed in a large number of chemical reactions. The main samarium catalysts described in the literature are iodide and oxide compounds, which can be used in a considerable number of different reactions.

Samarium iodides (SmI_2 and SmI_3) are efficient Lewis acids used in the reduction of azides or aromatic nitro compounds into the corresponding amines, in the formation of carbon-carbon bonds such as aldol, Michael, Diels-Alder and Michael-aldol reactions, Mannich and Makaiyama-aldol reactions amongst others. Samarium oxide, Sm_2O_3, is useful in methane coupling for example.

Herein, a review of samarium compounds applied to catalysis and their reactions is presented. This useful guide focus not only in reactions using samarium iodides and oxides as catalysts and pre-catalysts, but some other important and useful samarium catalysts such as $Sm(NO_3)_3$ in Pechmann reaction, $Sm(OTf)_3$ and other samarium halides ($SmCl_2$ and $SmBr_2$) in coupling reactions and relevant others.

Chapter 2 – Sm, a typical member of the lanthanide series, has many functional properties due to its internal particular 4f electronic structure. In particular, the luminescence and X-ray shielding of Sm have attracted considerable interest. In this chapter, the authors mainly focused on the recent development of Sm-complex/polymer composite because of its excellent functional properties, chemical stability and wide potential application. They also introduce a novel in-situ reaction to prepare the Sm-complex/polymer composite with fine dispersion and strong interfacial interaction. The luminescent composite exhibits a strong yellow color, high luminescent quantum and lifetime. These characteristics enable the composite with a broad application for flexible display, luminescent biosensor and laser systems. The X-ray shielding composite possesses less thickness and wide-range energy absorption compared with the classical Pb-contained composite, showing the practical application of human-body protection from X-rays leaking from a nuclear power station, security check, outer space, etc.

Chapter 3 – Samarium diiodide (SmI_2) is one of most important single-electron reducing agents in synthetic organic chemistry and commercially available as a SmI_2-THF solution. Samarium diiodide has the high redox potential and can reduce alkyl iodides, alkyl bromides, aldehydes, ketones, sulfoxides etc. However, the reduction with samarium diiodide itself often requires long period of heating. Therefore, several improved methods for enhancement of the reducing ability of samarium diiodide have been developed. For example, hexamethylphosphoric triamide (HMPA) as a Lewis base is the most effective activating reagent by forming a SmI_2-HMPA complex, $[SmI_2(hmpa)_4]$. Combination of SmI_2 and other metals such as typical metals, transition metals, and rare earth metals is also useful for the reduction of alkyl halides, carbonyl compounds, imines, acetates, and group 14 heteroatom chlorides. The addition of acids, bases, and/or proton sources also works to improve the reducing ability of SmI_2, and sometimes stereoselectivity. Moreover, the photoirradiated reduction system of SmI_2 has been developed as a powerful reduction method without addition of any additives. Currently, these activated SmI_2 systems are very important as synthetic methods in organic chemistry.

Chapter 4 – This paper reports on effective visible light induced reduction of Cr(VI) ion chiral Schiff base Cu(II) complex and TiO_2 in non-aqueous organic solvent (methanol) as a model system for environmental cleanup. By using several new and known chiral Schiff base Cu(II) complexes, hybrid systems of the Cu(II) complexes and anatase TiO_2 were prepared and reduction of Cr(VI) ion after visible light irradiation.

The Cu(II) complexes were synthesized in a common procedure and characterized by means of elemental analysis, IR, UV-vis, CD, and ESR spectra. Hybrid systems of the Cu(II) complex and TiO_2 were prepared as methanol suspension. After visible light irradiation, Cr(VI) ion was reduced successfully, which was confirmed by diphenylthiocarbazide analysis method. To our knowledge, this visible light reaction in non-aqueous organic solvents can be observed for the first time. However, reduction of Cr(VI) ion could not be observed for only a part of components selected of the hybrid systems. Although similar reaction was tested for Sm(III) ion instead of Cr(VI) ion for comparison, no change could be detected in emission as well as absorption spectra in the case of Sm(III) ion due to inappropriate redox potential.

Chapter 5 – Samarium has been studied extensively as a dopant in a wide number of host materials because of its chemical properties and the considerable modification on the properties of the host that are measured when even small concentrations of samarium are present. Samarium has the lowest activation energy of all of the rare earth elements. As a result, the ease at which charged vacancies in an oxide host material such as cerium oxide (ceria) are generated is observed as an increase in the ionic mobility and conductivity of the oxide host. In addition, an increase in the fluorescence intensity has been measured in samarium-doped rare earth oxides when compared to the undoped oxides. The absorbance dispersion of samarium-doped materials exhibits a blue shift into the ultra violet as well. Following a description of the characteristics of samarium-doped hosts, which will include discussion the reasons for the chemical, optical, and structural properties that result from the presence of samarium, a review of the applications of the samarium -doped host materials will be presented. These include environmental sensing via electrochemical and fluorescence quenching techniques as well as their potential use as phosphors in solid state lighting and as a nanopharmaceutical diagnostic or therapeutic agent.

In: Samarium
Editor: Kaitlyn R. Danford

ISBN: 978-1-63321-045-5
© 2014 Nova Science Publishers, Inc.

Chapter 1

SAMARIUM COMPOUNDS: AMAZING WORLD OF CATALYSTS

Cláudia D. Raposo[*]
REQUIMTE, CQFB, Departamento de Química,
Faculdade de Ciências e Tecnologia da Universidade Nova de Lisboa,
Caparica, Portugal

ABSTRACT

In catalysis, a relatively small amount of a material (named catalyst) is added to increase the rate of the reaction, without it being consumed, thus allowing the speed up of a reaction and the selective formation of a product. Catalysts are very important in chemistry and are used in more than 80 % of industrial processes.

Samarium is a lanthanide that preferentially exists in the +3 oxidation state. Samarium compounds are useful as catalysts which have been employed in a large number of chemical reactions. The main samarium catalysts described in the literature are iodide and oxide compounds, which can be used in a considerable number of different reactions.

Samarium iodides (SmI_2 and SmI_3) are efficient Lewis acids used in the reduction of azides or aromatic nitro compounds into the corresponding amines, [1, 2] in the formation of carbon-carbon bonds such as aldol, Michael, Diels-Alder and Michael-aldol reactions [3],

---

[*] Email: piccfa@gmail.com.

Mannich and Makaiyama-aldol reactions [4] amongst others. Samarium oxide, Sm_2O_3, is useful in methane coupling for example [5].

Herein, a review of samarium compounds applied to catalysis and their reactions is presented. This useful guide focus not only in reactions using samarium iodides and oxides as catalysts and pre-catalysts, but some other important and useful samarium catalysts such as $Sm(NO_3)_3$ in Pechmann reaction, $Sm(OTf)_3$ and other samarium halides ($SmCl_2$ and $SmBr_2$) in coupling reactions [6,7,8,9] and relevant others.

1. INTRODUCTION

In catalysis, a relatively small amount of a material – the catalyst – is added to increase the rate of the reaction, without it being consumed. The main goal is to speed up the chemical reaction and often to obtain selectively the desired product, producing less waste in a smooth reaction. A catalyst forms bonds with the reacting molecules, allowing them to react to a product which detaches from the catalyst and leaves it unaltered and available for the next reaction. Therefore the catalytic reaction is a cyclic event [10].

The catalyst offers a more energetically favorable path for the reaction, thus accelerating both the forward and the reverse reaction to the same extent. The activation energy of the catalytic reaction is smaller, thus the rate of the catalytic reaction is much larger. It is important to note that if the bonding between reactants and catalyst is too weak, the reaction will hardly occur. On the other hand, if the bond between the catalyst and one of the reactants is too strong, then the product will hardly be formed. If the product strongly bonds to the catalyst, than it is very difficult for separation to occur and the product poisons the catalyst [10].

Catalysts are very important in chemistry and are used in more than 80 % of industrial processes. Since chemical processes transform raw materials into products, it is important that these reactions are relatively quick and selective so they can be considered for an industrial scale [10].

Different combinations of ligands and metals are used to perform highly enantioselective reactions. Transition metal catalysts are often moisture and air-sensitive or are present as contaminants in products. Organocatalysis uses small organic molecules which are stable towards moisture or air as the catalytic active species. Whereas transition metal catalysts interact with substrates by coordination, organocatalysts offer the possibility to interact either by forming reactive covalent intermediates (enamines or imines for instance) or by forming hydrogen-bond or ion-pair complexes [11].

Samarium is a lanthanide that preferentially exists in the +3 oxidation state. Samarium compounds have been used as useful catalysts in a large number of chemical reactions. The main samarium catalysts described in the literature are iodide and oxide compounds, which can be used in a considerable number of different reactions.

Samarium iodides (SmI_2 and SmI_3 for instance) being efficient Lewis acids, are used in the reduction of azides or aromatic nitro compounds into the corresponding amines [1,2], in the formation of carbon-carbon bonds such as aldol, Michael, Diels-Alder, [3] Mannich and Makaiyama-aldol reactions[4] amongst others. Samarium oxide, Sm_2O_3 is useful in methane coupling for example [5].

This guide seeks to provide a review of main samarium compounds applied to catalysis and their reactions, based on research results available in the literature. Samarium iodides and oxides are the main samarium compounds used as catalysts or even pre-catalysts, but some other important and useful samarium catalysts such as $Sm(NO_3)_3$ in Pechmann condensation, $Sm(OTf)_3$ and samarium halides ($SmCl_2$ and $SmBr_2$) in coupling reactions [6,7,8,9] and relevant others will be presented.

2. SAMARIUM HALIDES

2.1. Samarium Diiodide (SmI_2)

Samarium(II) diiodide (also known as diiodosamarium), SmI_2, can be a versatile compound since it can be used in a large variety of reactions acting either as the active catalyst or as pre-catalyst. It is commercially available both as an anhydrous powder and as a solution in THF, but can also be prepared readily using one of the various methods that have been described. [12],[13],[14] The preparation of SmI_2 is very often, since this compound is air sensitive (although somewhat tolerant of water). Samarium diiodide can be inactivated even after a brief exposure to air and/or moisture, and its preparation is both cheap and easy. This catalyst is stable under nitrogen or argon.

Samarium diiodide has been utilized in a wide range of synthetic transformations since it can donate one electron (Sm^{3+} is the stable oxidation state), acting as an efficient Lewis acid-type catalyst for a variety of carbon-carbon and carbon-nitrogen bond-forming reactions such as Michael reactions,

Diels-Alder reactions, tandem Michael-aldol reactions, [3] Meerwein-Poondorf-Verley reductions, [15] amongst others as it will be described next.

It is notable the ability of SmI_2 to modify its behaviour in the presence of cosolvents or additives which can affect the rate, the yield, the chemo- or stereoselectivity of the reaction. Additives commonly used are Lewis bases like hexamethylphosphoramide (HMPA), proton sources like alcohols and water and inorganic additives like nickel(II) iodide (NiI_2) and iron(III) chloride ($FeCl_3$) [16].

Beneficial influence of HMPA as an additive was noticed on some reactions mediated by SmI_2. Samarium Barbier reactions in THF, for instance, are much faster in the presence of HMPA and allowing the use of room temperature instead of reflux conditions. *N,N'*-Dimethylpropylene urea (DMPU) is less toxic than HMPA and therefore can replace it in some reactions. 1,1,3,3-tetramethylurea (TMU) has also been employed [16,17].

Solvents have a crucial influence in samarium(II) diiodide chemistry since some reactions can be more selective, accelerated and some organosamarium species can be stabilized. Protic additives such as methanol or *tert*-buthanol are often essential to the development of the desired reaction. This is because *in situ* protonation of key intermediates or end products occur and therefore the great difference between product distribution in protic conditions compared with aprotic conditions. Water can sometimes be added to accelerate some reductions [16].

One severe limitation to the samarium(II) diiodide chemistry is the samarium metal necessary for preparation of SmI_2, which can be quite expensive and therefore the use of equimolar amounts of samarium(II) diiodide is not praticable. It is desirable to reduce the amount of inorganic reagent by devising a catalytic process which can be achieved by using magnesium, zinc or other metal salts as the co-reducing agent [16,17,18].

A system for *in situ* regeneration of samarium(II) diiodide should be cheaper than samarium metal, able to quickly reduce samarium(III) species into samarium(II) ones, unreactive toward present substrates and simple and usable in most of SmI_2-mediated organic reactions [19].

2.1.1. Diels-Alder Reactions

Imino-Diels-Alder Reactions

Cycloaddition reactions of imines with various dienes or alkenes were performed by Collin, Jaber and Lannou in the presence of tetrahydrofuran solvated samarium diiodide at room temperature and the desired products were obtained in high yields (figure 1) [3].

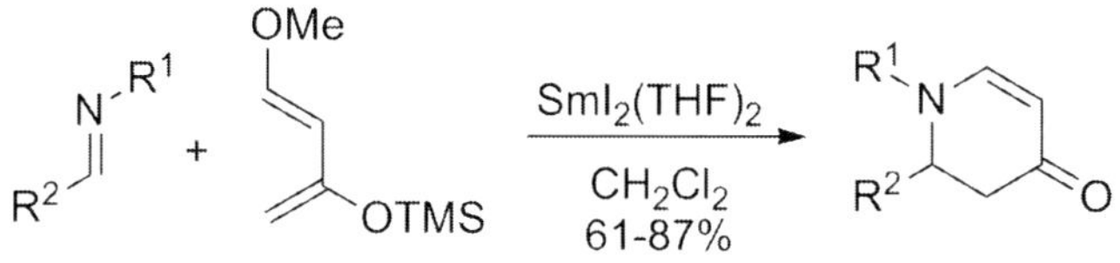

Figure 1. Cycloaddition of imines with Danishefsky's diene [3].

2.1.2. Aldol, Michael and Tandem Aldol-Michael Reactions

Imino-Aldol Reactions

Collin and co-workers stated that samarium(II) iodide also catalyzes imino-aldol reactions with cyclic enoxysilanes leading to β-amino ketones in good to excellent yields (figure 2) [3].

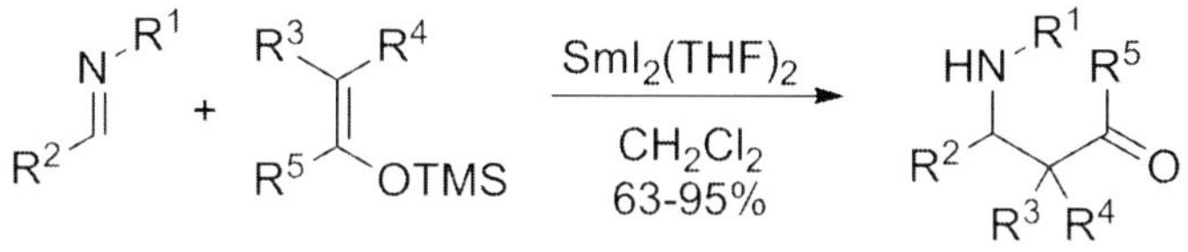

Figure 2. Imino-aldol reactions catalyzed by samarium(II) iodide [3].

2.1.3. Barbier Reactions

Another class of transformations of wide interest is the Barbier type reaction. Kagan mixed equimolar amounts of a ketone and an organic halide (RX) with two equivalents of SmI_2 in a one-pot procedure (figure 3) [16],[17].

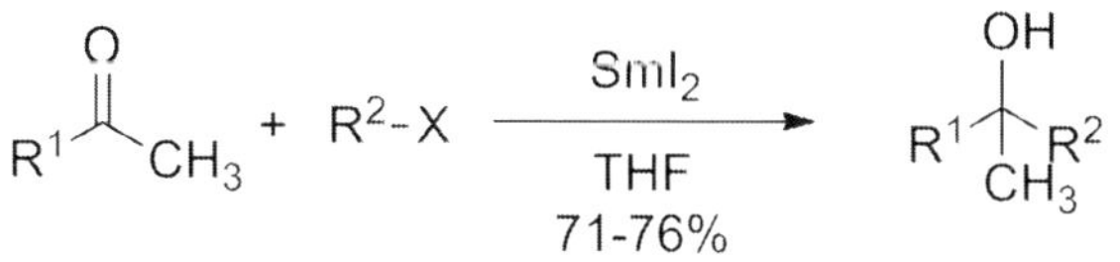

Figure 3. Samarium-Barbier reaction [16].

In the Barbier reaction is possible to use a variety of allylic and benzylic halides, which are prone to give self-coupling products in the presence of SmI_2 but are also excellent reactants in the Barbier reaction [16].

Intramolecular samarium Barbier reaction is also possible and it is interesting that ofthen the cyclization is highly diastereoselective. Polyquinanes have been obtained by a bis-Barbier reaction [17].

Hélion and Namy reported the use of mischmetall in samarium(II) diiodide-catalyzed Barbier reactions. It was possible to use catalytic amounts of samarium diiodide in the presence of a mischmetall suspension [19].

2.1.4. Reformatsky Reactions

Reformatsky reaction mediated by SmI_2 provides a useful alternative to traditional versions of the reaction since it proceeds under mild and homogeneous conditions, with high chemo- and diastereoselectivity.

The samarium-Reformatsky reaction was found to occur very easily using 2 equivalents of samarium(II) diiodide as the promotor (figure 4).

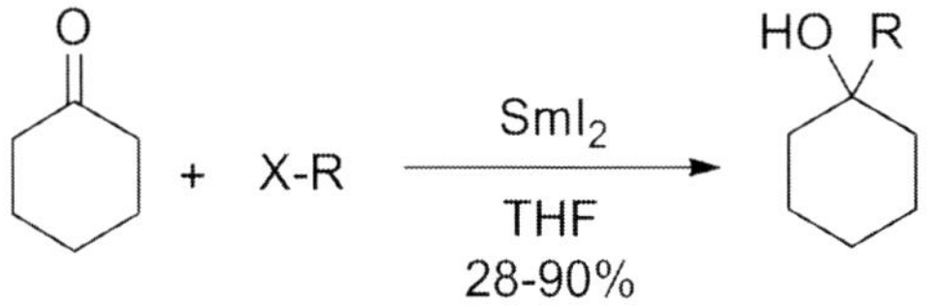

Figure 4. Samarium-Reformatsky reaction [16].

Samarium-Reformatsky reactions of chiral 3-bromoacetyl-2-oxozolidinones have been described by Fukuzawa. The product was obtained in good yield (87%) and high diastereoisomeric excess (99%). The reaction is noteworthy since highly diastereoselective acetate aldol processes are difficult to achieve [20].

Otaka presented a SmI_2-mediated Reformatsky reaction involving formaldehyde as the electrophile in a synthesis of a dipeptide isostere in good yield (65%) [21].

Reformatsky cyclisations using SmI_2 were also used by Mukaiyama in total synthesis of paclitaxel (Taxol) [22].

2.1.5. Pinacol Coupling Reactions

Since the discovery of pinacol coupling reaction by Wilhelm Fittig, this reaction has become an important synthetic tool for the construction of carbon-carbon bonds.

Samarium(II) diiodide catalyzes carbonyl couplings as summarized in figure 5 [16].

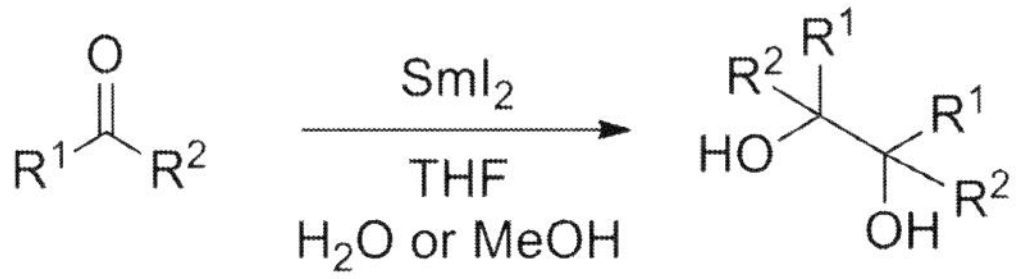

Figure 5. Carbonyl coupling catalyzed by SmI_2.

Intermolecular reductive dimerisation of aldehydes or ketones gives dymmetrical diols. In aprotic conditions, aldehydes and ketones can therefore generate pinacols. The reaction is very fast with aromatic aldehydes or ketones and can be highly chemoselective. 16] Arylaldehydes and aryl ketones couple within seconds at room temperature while aliphatic aldehydes and ketones react more slowly (several hours to complete the reaction). Addition of HMPA can greatly accelerate these slower coupling reactions [23].

Pinacol coupling reaction in the presence of mischmetall suspension have also been reported by Helion and Namy [19].

2.1.6. Reduction of Alkyl Halides

Samarium(II) diiodide can reduce alkyl halides to the corresponding alkanes with high yields and with the ease of reduction as it follows: iodides > bromides > chlorides. The addition of HMPA to SmI_2 allows milder conditions, reducing alkyl bromides at room temperatures and alkyl chlorides upon heating. It is important to note that these reductions can be carried out in the presence of other functional groups [23,24,25,26].

The reduction of alkyl halides with samarium(II) diiodide has been used to prepare alkenes by β-elimination reactions. In the presence of more than one halide in the molecule (different halides), only the more reactive halide is reduced [27,28,29].

2.1.7. Reduction of Ketones and Aldehydes

SmI_2 reduces ketones and aldehydes to the corresponding alcohols. Although there are numerous hydride reducing agents, in some cases SmI_2 can be more reactive, stereoselective and chemoselective and thus its importance.

Reduction of ketones and aldehydes can be acomplished using THF as solvent and in the presence of two equivalents of samarium(II) diiodide. Aldehydes and ketones are therefore reduced into the corresponding alcohols as summarized in figure 6 [16].

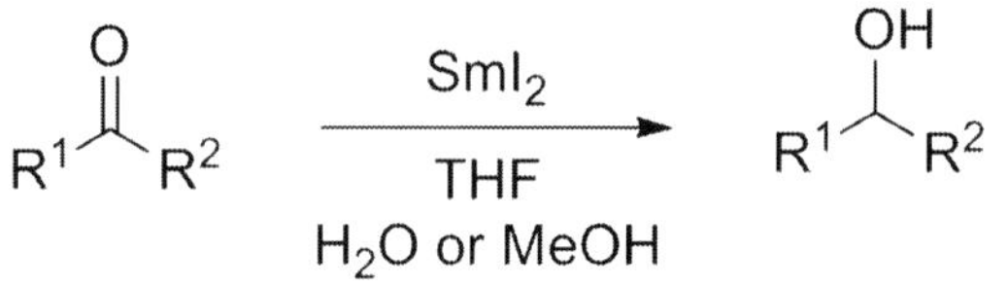

Figure 6. Carbonyl reduction catalyzed by SmI_2 [16].

Intramolecular Tishchenko reduction of β-hydroxy ketones has been studied by Evans and Hoveyda. [30] This reaction affords the corresponding anti-diol monoesters in high yields and with excellent levels of stereochemical control (figure 7).

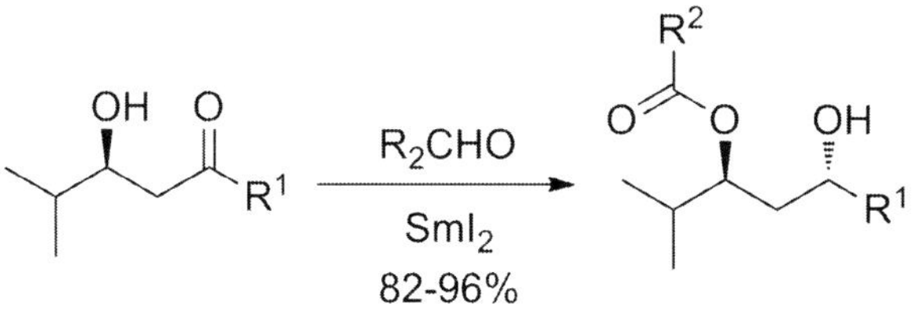

Figure 7. Tishchenko reaction catalyzed by samarium(II) diiodide [30].

The active samarium catalyst can be a samarium(III) compound formed during the course of reaction since Evans and Hoveyda observed the change of the blue color of samarium(II) to yellow-orange color of samarium(III) upon addiction of SmI_2 to the reaction mixture [30].

Mascarenhas et al., reported a stereoselective hetero-aldol-Tishchenko reaction (figure 8) using a samarium alkoxide as catalyst. [31] Coupling an irreversible Tishchenko reaction to a reversible aldol reaction gives rise to a catalytic aldol-Tishchenko reaction with high product yields (aldol process), not requiring stoichiometric preactivation of initial ketone as a metal enolate. Mascarenhas observed alkali metal and alkaline earth alkoxides tended to be most efficient out of those alkoxides examined [31].

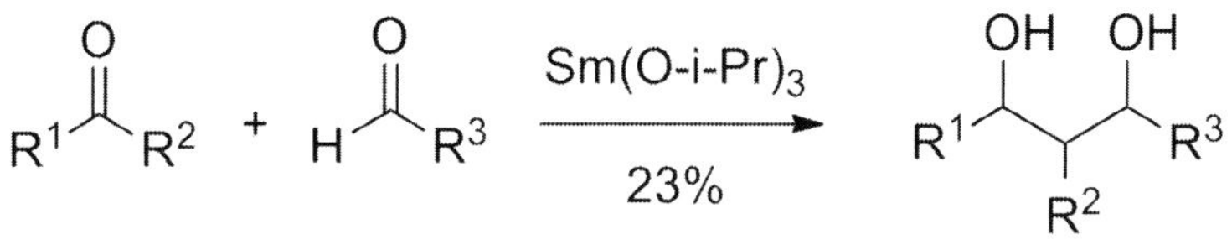

Figure 8. Catalytic hetero-aldol-tishchenko reaction.

2.1.8. Reduction of Carboxylic Acids and Esters

If an activating additive is present, reducing of simple unfunctionalised substrates is possible using SmI_2. Kudo reported the reduction of carboxylic acids with samarium(II) diiodide, water and sodium hydroxide. The use of water as cosolvent is known to activate SmI_2 and its inclusion is crucial to the success of these reactions (figure 9) [32].

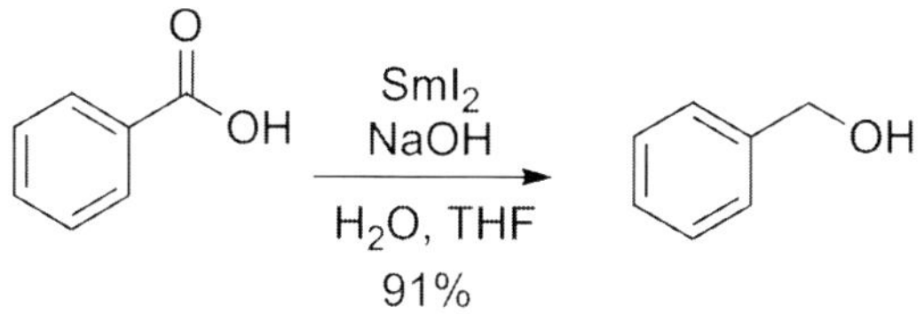

Figure 9. Reduction of carboxylic acids promoted by SmI_2 [32].

Marko studied the reduction of primary, secondary and teriaty toluates using SmI_2 and HMPA affording the corresponding deoxygenated products in good yield (figure 10) [33].

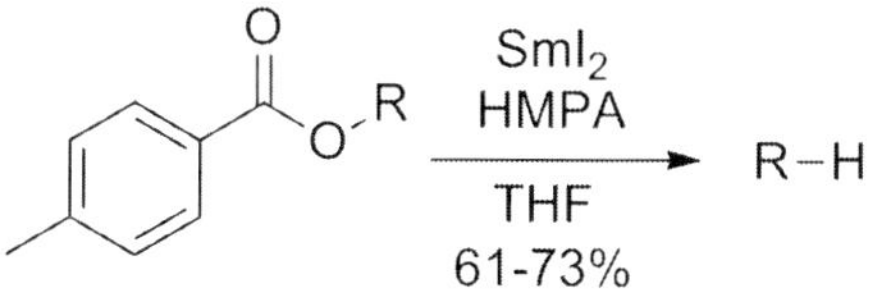

Figure 10. Reduction of esters promoted by SmI_2-HMPA [33].

2.1.9. Reduction of Aromatic Nitro Compounds to Aromatic Amines

A simple method for the reduction of aromatic nitro compounds to corresponding aromatic amines was presented by Banik and co-workers using samarium and catalytic amounts of iodine, thus preparing SmI_2 *in situ* (figure 11) in good to excellent yields. [2] This method can be chosen over catalytic hydrogenation if the starting material needs to be handled with care.

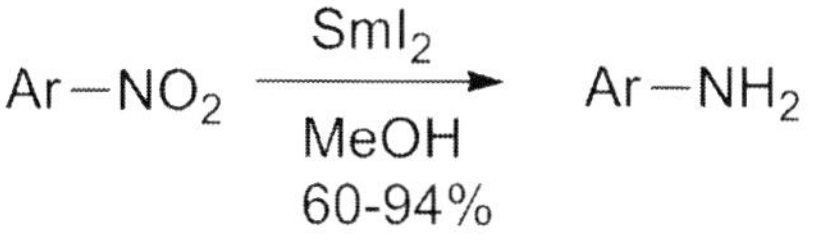

Figure 11. Reduction of aromatic nitro compounds into corresponding amines [2].

The reduction process showed some selectivity since there was no dehalogenation in the reduction of 4-bromonitrobenzene to 4-bromoaniline neither hydrogenolysis in the reduction of 2-nitro-9-ketofluorene to 2-amino-9-ketofluorene. It is noteworthy that catalytic hydrogenation is not that selective, leading to dehalogenation and hydrogenolysis [2].

2.1.10. Preparation of α-Amino Phosphonates

Widely used in biochemical and pharmaceutical chemistry, a-amino phosphonates are an important class of compounds which synthesis has been the subject of much interest. Samarium(II) diiodide was found to catalyze C-N bond-forming reactions and Xu et al., carried out the reaction of benzaldehyde with aniline and diethyl phosphite in the presence of 10 mol % SmI_2. The reaction proceeded smoothly with good yields (figure 12)[34].

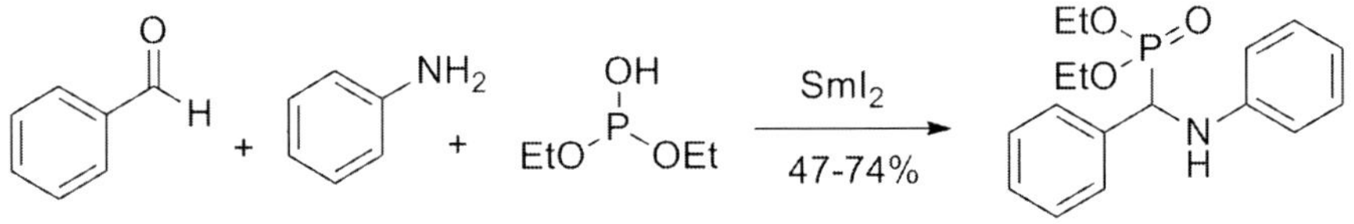

Figure 12. Samarium diiodide used as pre-catalyst for the preparation of α-amino phosphonates [34].

The change in the color of the reaction mixture from dark blue to yellow and the formation of the solid imine suggests that samarium(III) species, supposedly the active species for this reaction, is generated during the reaction of SmI_2 with aldehyde or amine [34].

2.2. Samarium Triiodide (SmI₃)

2.2.1. Michael Reaction

The reaction of indoles with electron-deficient olefins has been studied by Zhan, leading to the corresponding Michael adducts in high yields. [35] The simple and direct method described for the synthesis of 3-alkylated indoles consists in the conjugate addition of indoles to α,β-unsaturated compounds in the presence of the Lewis acid SmI_3 (figure 13). Since the addition of indoles requires careful control of acidity to prevent side reactions (namely dimerization and polymerization), samarium(III) triiodide is an efficient Lewis acid catalyst [35].

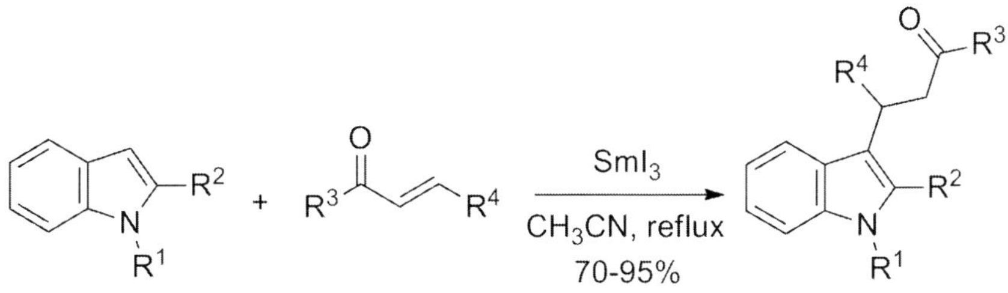

Figure 13. Michael reaction catalyzed by samarium(III) triiodide [35].

The reactions performed by Zhan and co-workers were clean, leading to the desired products in high yields without the formation of side products (such as *N*-alkylation product). The indole nitrogen did not require prior protection thus permiting a wide range of functional groups to be used. Plus, the procedure did not require any acidic promoters or inert atmospheric condition [35].

2.2.2. Hydroboration of Olefins

Evans and co-workers studied lanthanide hydroboration catalysis to prepare boronate esters. A set of trivalent samarium catalysts were screened. SmI_3 along with $(t\text{-BuO})SmI_2$, and $(i\text{-PrO})_3Sm$ were found to be competent catalysts affording the desired product in good yields [36].

2.3. Samarium Chlorides ($SmCl_2$ and $SmCl_3$)

Shiraishi et al., found that both SmI_2 and $SmCl_3$ catalyze the three-component coupling of amines, α,β-aldehydes or ketones and nitroalkanes to give the corresponding pyrroles in fair yields (55-64%, figure 14) [37].

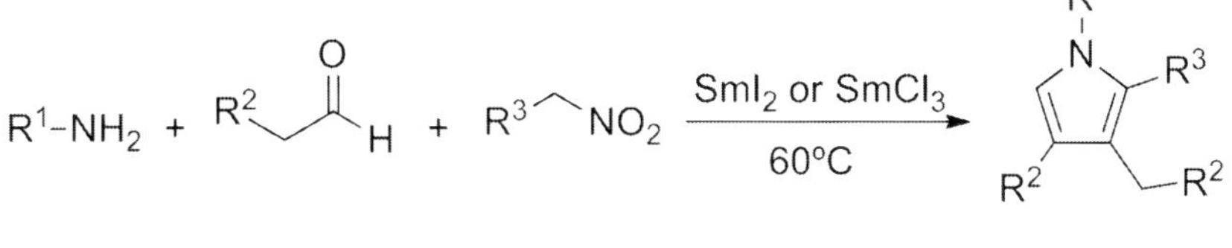

Figure 14. Pyrrole preparation using samarium-based catalysts [37].

Samarium(II) chloride can be prepared *in situ* for pinacol coupling reactions and can also be used as a critical component in the intramolecular coupling of cyanates and α,β-unsaturated cyclic ketones to produce spirooxindoles [7].

2.4. Samarium Bromide ($SmBr_2$)

The main advantage of using samarium(II) bromide over samarium(II) iodide is that it is a very efficient reagent for the pinacol coupling of aldehydes and ketones and, in some cases, cross coupling between different carbonyl compounds is efficient [6].

3. SAMARIUM OXIDES

3.1. Catalytic Incineration by $SmMnO_3$

The process of incineration is the most often method used to control emission to the atmosphere of volatile organic compounds (VOCs) from process industries.

Catalytic incineration destroys the pollutant molecules as an alternative route to the typical thermal process and therefore saving energy. Noble metals are usually preferred since they generally allow lower operating temperatires and higher space velocities than the transition metal oxide based catalysts. However, transition metal oxides are not that expensive and often noble metals lead to sublimation and sinterization problems [38].

Catalytic combustion of methane has been widely investigated. [39,40] Since catalytic materials to be used in the combuster must have high surface area, thermal stability, high durability with respect to oxidation activity and product selectivity, transition metal oxides can be used as perovskite-type oxides (general formula ABO_3, where A and B are usually rare earth and transition metal cations, respectively).

The effect of temperature on the activity of manganese-based catalysts in the complete oxidation of acetone has been studied by Gil et al.,[38] It has been observed a positive effect for the samarium-manganese-based catalyst ($SmMnO_3$) performance with respect to the single manganese ones. Due to their high performance, the oxides studied can be considered suitable catalysts for high temperature processes [38].

3.2. Esterification Catalyzed by Samarium Oxide Nanoparticles

Nanoparticles can be employd in catalysis since larger specific surface area and higher activity of the nanoparticles can improve catalytic efficiency dramatically. Nanometer-sized Sm_2O_3 poweders has been syntesized by Gao et al., thus obtaining homogeneous spherical particles. The catalytic activity for the esterification of phtalic anhydride with 2-ethyl hexanol was tested (figure 15) and the results showed the nanoscale Sm_2O_3 could catalyze esterification more efficiently than the macroscopic counterpart. Reduced reaction time and improvement of reaction yield was observed [35].

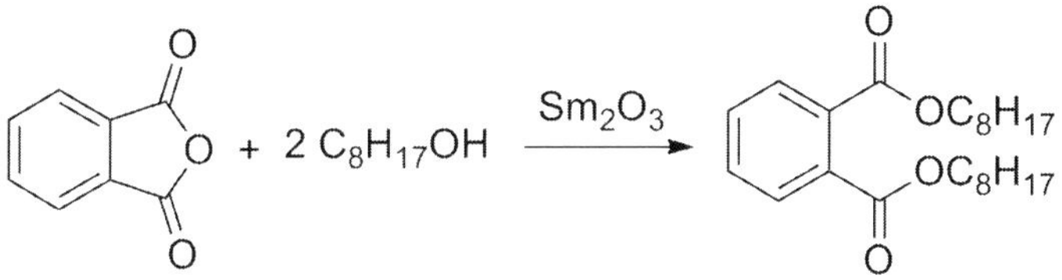

Figure 15. Catalyzed esterification using nanoscale Sm_2O_3 powder [35].

3.3. Methane Coupling (Sm_2O_3)

The oxidative coupling of methane usually proceeds by dimerization of methyl radicals produced at the surface of a catalyst, usually an oxide. The reaction seems to occur by promotion of lanthanide oxides as catalysts. Capitán et al., studied methane coupling in the presence of a system of mixed oxides, namely Sm_2O_3 and Al_2O_3 with good results [5].

4. Nitrogen-Containing-Samarium Compounds

4.1. Pechman Reaction

The Pechmann reaction is largely applied for the preparation of coumarins. Coumarins are an important class of both natural and synthetic products which have been used as additives (food and cosmetics), anticoagulants, in the preparation of insecticides, laser dyes, etc.

The construction of coumarins involves the condensation of phenols with β-ketonic esters in the presence of a variety of acidic condensing agents with good yields. The acidic condensing agents are often used in excess which leads to environmental pollution due to disposal of acidic waste. Bahekar and Shinde reported a general and practical route for the Pechmann condensation using samarium(III) nitrate hexahydrate as the catalyst under solvent-free conditions (figure 16) [41].

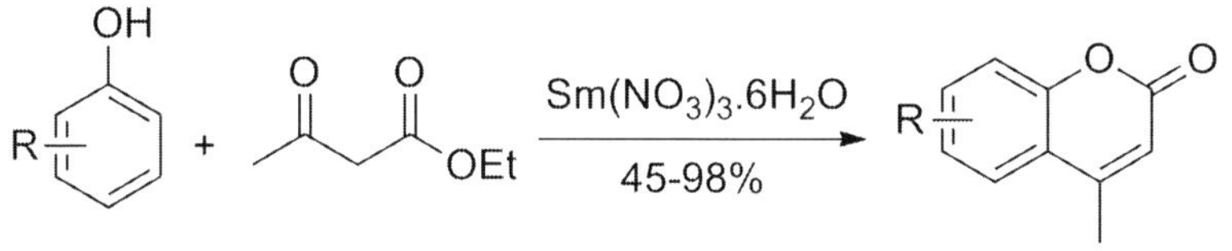

Figure 16. Synthesis of coumarins via von Pechmann condensation catalyzed by $Sm(NO_3)_3.6H_2O$ [41].

5. OTHER RELEVANT SAMARIUM COMPOUNDS

5.1. Samarium Trifluoromethanesulfonate ($Sm(OTf)_3$)

Samarium trifluoromethanesulfonate (a kind of Lewis acid) is an effective catalyst for transesterification of dimethyl carbonate with phenol to synthesize diphenyl carbonate (figure 17) which is an industrially important intermediate for the production of a lot of organic chemicals [42].

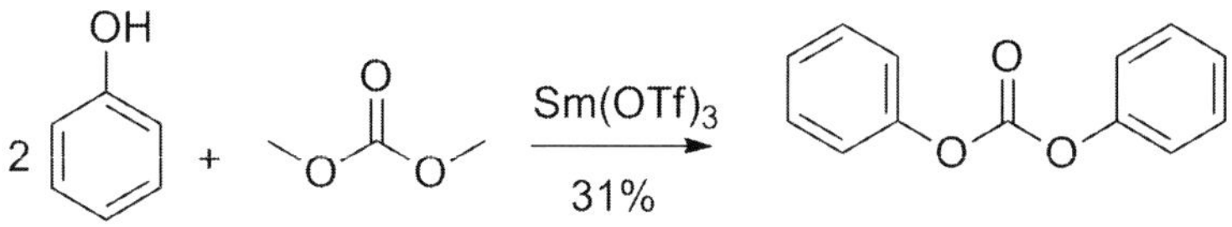

Figure 17. Transesterification catalyzed by samarium trifluoromethanesulfonate [42].

5.2. Chiral Samarium-Based Catalyst

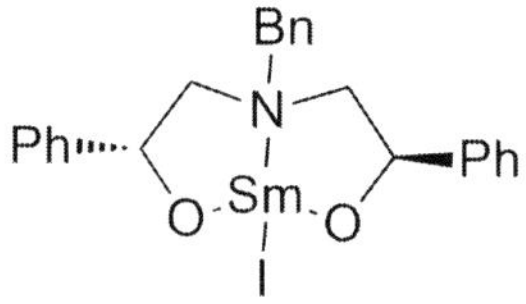

Figure 18. Chiral samarium(III)-based catalyst.

A chiral samarium(III) complex has been used at ambient temperatures for the Meerwein-Ponndorf-Verley reaction for the reduction of ketones. Evans and co-workers synthesized the catalyst (figure 18) which was subsequently used to catalyze selectively the reduction of ketones in excellent yield and enantiomeric excess (figure 19) [15].

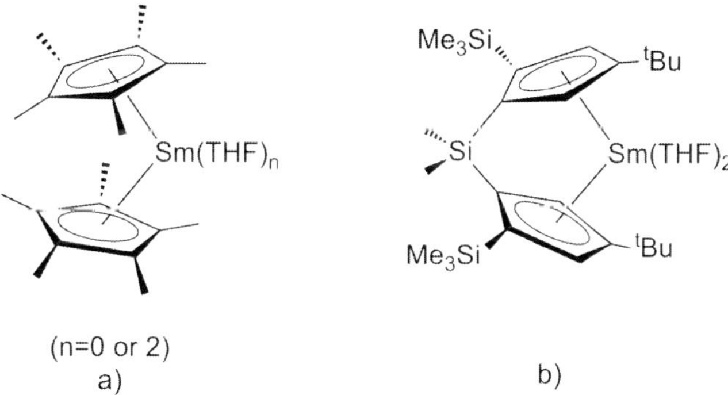

Figure 19. Assimetric reduction of ketones [15].

They studied other lanthanide(III) complexes and found that the most active lanthanide catalysts derived from the methal halide (Y, Tb, Sm and Nd). Also, reduction times were observed thus concluding the complex derived from samarium(III) to be the optimal catalyst on both selectivity and reactivity considerations [15].

5.3. Divalent Samarium Metallocene Complexes

Figure 20. Samarium di-metallocene complexes used as catalysts.

Development of more efficient or selective catalysts can by achieved by changing the ligand environment of a metal complex to modify its properties.

Some lanthanide-based catalysts display advantages over analogous d-block transition metal-based catalysts (higher activity and stereoselectivity, good living character in polymerization and no requirement for a cocatalyst or activator). Divalent samarium metallocene complexes (figure 20) have received much interest both as homogeneous catalysts and precatalysts for other transformations [43].

Polymerization of styrene, ethylene (and copolymerization of both) by lanthanide metallocene catalytic system has been studied by Hou et al., with good results [43].

Boffa and Novak also studied bifunctional polymerization initiators, namely figure 20.a) type of catalyst and the well-controlled polymerization of methyl methacrylate with the described bimetallic samarium catalyst proceeded with promising results [44, 45].

5.4. Samarium Iodobinaptholates

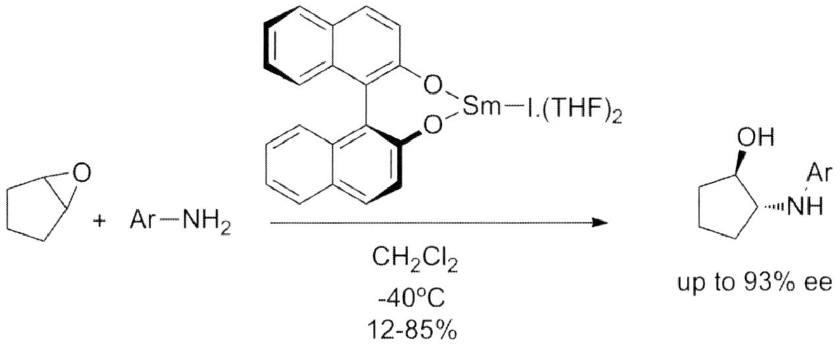

Figure 21. Samarium iodobinaphthoxide catalyzes enantioselective aminolysis of cyclic *meso*-epoxides [46].

Samarium iodobinaphtholates are efficient catalysts for the enantioselective ring opening of *meso*-epoxides by aromatic amines as observed by Carrée et al. (figure 21). [46] Catalyzed enantioselective ring opening of epoxides with amines affords an easy route to asymmetric β-amino alcohols. Carrée and co-workers observed good to high asymmetric inductions with enantiomeric excesses up to 93%. They also observed that acyclic substrates were less reactive and samarium catalyst is more active and enantioselective than other catalysts used in similar reactions [46].

Bis-binaphthoxy iodo samarium catalyzes enantioselectively Mannich-type reactions aswell. The catalytic reaction for the addition of ketene silyl acetal on a glyoxylic imine has been studied by Jaber et al., and the reaction proceeded smoothly in mild conditions with good to excellent yields (figure 22) [4].

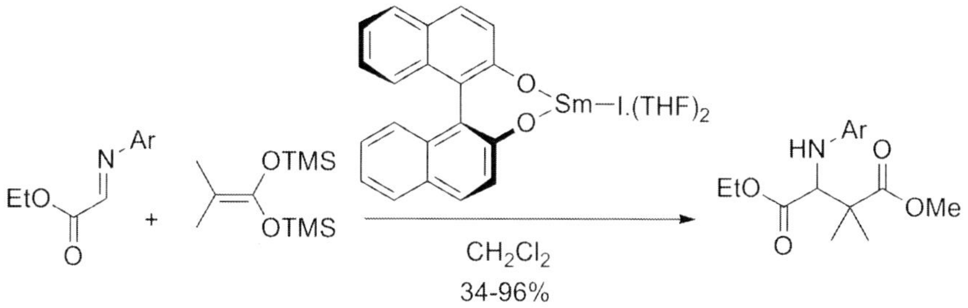

Figure 22. Mannich reaction catalyzed by samarium iodobinaphthoxide [4].

CONCLUSION

Although samarium-catalysts are best known for the powerfull carbon-carbon bond-forming reactions, these transformations are initiated by the reduction of functional groups. The ability to manipulate a particular functional group in the presence of others is an important characteristic and samarium catalysts are the reagent to be chosen.

Unfortunately, in catalytic SmI_2 reagent systems, iodide ligands are displaced through coordination to intermediates and as a result, the samarium(III) species reduced in the catalytic cycle may produce a reductant other than samarium(II) diiodide.

Samarium(II) diiodide is as valuable precursor to other samarium(II) and (III). The most common method for generating new species is by using additives and cosolvents.

Cosolvents and additives can have a positive and remarkable effect of samarium-mediated reactions, namely an improvement in reaction rate, yield or stereoselectivity or even more dramatic enabling of an otherwise impossible reaction.

SmI_2-promoted pinacol couplings are one of the most useful reactions for the stereoselective preparation of 1,2-diols in acyclic and cyclic systems. Extension of these reactions to asymmetric synthesis of β-amino alcohols and vicinal diamines through the coupling of a vatiety of imine derivatives represents a significant advantage in organic chemistry. Samarium(III) enolates are important intermediates in many samarium diiodide reactions. The reduction of α-halo-substituted carbonyl compounds with SmI_2 is the most used method for the preparation of Sm(III) enolates. Reformatsky and aldol reactions promoted by samarium can be used in inter and intramolecular reactions and have many advantages over other metal-mediated reactions due to the high reaction regio-, chemo- and diastereoselectivity.

REFERENCES

[1] Y. Huang, Y. Zhang, Y. Wang, *Tetrahedron Letters* 1997, *38*, 1065-1066.

[2] B. Banik, C. Mukhopadhyay, M. Venkatraman, F. Becker, *Tetrahedron Letters* 1998, *39*, 7243-7246.

[3] J. Collin, N. Jaber, M. Lannou, *Tetrahedron Letters* 2001, *42*, 7405-7407.

[4] N. Jaber, F. Carree, J. Fiaud, J. Collin, *Tetrahedron-Asymmetry* 2003, *14*, 2067-2071.

[5] M. Capitan, P. Malet, M. Centeno, A. Munozpaez, I. Carrizosa, J. Odriozola, *Journal of Physical Chemistry* 1993, *97*, 9233-9240.

[6] A. Lebrun, J. Namy, H. Kagan, *Tetrahedron Letters* 1993, *34*, 2311-2314.

[7] S. Reisman, J. Ready, M. Weiss, A. Hasuoka, M. Hirata, K. Tamaki, T. Ovaska, C. Smith, J. Wood, *Journal of the American Chemical Society* 2008, *130*, 2087-2100.

[8] S. Fukuzawa, K. Mutoh, T. Tsuchimoto, T. Hiyama, *Journal of Organic Chemistry* 1996, *61*, 5400-5405.

[9] S. Fukuzawa, Y. Yahara, A. Kamiyama, M. Hara, S. Kikuchi, *Organic Letters* 2005, *7*, 5809-5812.

[10] I. Chorkendorff, J. W. Niemantsverdriet, *Concepts of modern catalysis and kinetics*, Wiley-VCH, Weinheim ; [Cambridge], 2003.

[11] M. Rueping, R. Koenigs, I. Atodiresei, *Chemistry-a European Journal* 2010, *16*, 9350-9365.

[12] P. Girard, J. L. Namy, H. B. Kagan, *Journal of the American Chemical Society* 1980, *102*, 2693-2698.

[13] T. Imamoto, M. Ono, *Chemistry Letters* 1987, 501-502.

[14] J. M. Concellon, H. Rodriguez-Solla, E. Bardales, M. Huerta, *European Journal of Organic Chemistry* 2003, 1775-1778.

[15] D. A. Evans, S. G. Nelson, M. R. Gagne, A. R. Muci, *Journal of the American Chemical Society* 1993, *115*, 9800-9801.

[16] H. Kagan, *Tetrahedron* 2003, *59*, 10351-10372.

[17] H. Kagan, *Journal of Alloys and Compounds* 2006, *408*, 421-426.

[18] E. Corey, G. Zheng, *Tetrahedron Letters* 1997, *38*, 2045-2048.

[19] F. Helion, J. Namy, *Journal of Organic Chemistry* 1999, *64*, 2944-2946.

[20] S. Fukuzawa, H. Matsuzawa, S. Yoshimitsu, *Journal of Organic Chemistry* 2000, *65*, 1702-1706.

[21] A. Otaka, J. Watanabe, A. Yukimasa, Y. Sasaki, H. Watanabe, T. Kinoshita, S. Oishi, H. Tamamura, N. Fujii, *Journal of Organic Chemistry* 2004, *69*, 1634-1645.

[22] T. Mukaiyama, I. Shiina, H. Iwadare, M. Saitoh, T. Nishimura, N. Ohkawa, H. Sakoh, K. Nishimura, Y. Tani, M. Hasegawa, K. Yamada, K. Saitoh, *Chemistry-a European Journal* 1999, *5*, 121-161.

[23] J. Inanaga, M. Ishikawa, M. Yamaguchi, *Chemistry Letters* 1987, 1485-1486.

[24] H. B. Kagan, J. L. Namy, P. Girard, *Tetrahedron* 1981, *37*, 175-180.

[25] E. Hasegawa, D. P. Curran, *Tetrahedron Letters* 1993, *34*, 1717-1720.

[26] D. P. Curran, T. L. Fevig, C. P. Jasperse, M. J. Totleben, *Synlett* 1992, 943-961.

[27] J. M. Concellon, H. Rodriguez-Solla, *Chemical Society Reviews* 2004, *33*, 599-609.

[28] J. M. Concellon, P. L. Bernad, J. A. Perez-Andres, *Angewandte Chemie-International Edition* 1999, *38*, 2384-2386.

[29] J. M. Concellon, P. L. Bernad, E. Bardales, *Organic Letters* 2001, *3*, 937-939.

[30] D. Evans, A. Hoveyda, *Journal of the American Chemical Society* 1990, *112*, 6447-6449.

[31] C. M. Mascarenhas, M. O. Duffey, S. Y. Liu, J. P. Morken, *Organic Letters* 1999, *1*, 1427-1429.

[32] Y. Kamochi, T. Kudo, *Chemistry Letters* 1991, 893-896.

[33] K. Lam, I. E. Marko, *Organic Letters* 2008, *10*, 2773-2776.

[34] F. Xu, Y. Luo, M. Deng, Q. Shen, *European Journal of Organic Chemistry* 2003, 4728-4730.

[35] Z. P. Zhan, R. F. Yang, K. Lang, *Tetrahedron Letters* 2005, *46*, 3859-3862.

[36] D. A. Evans, A. R. Muci, R. Sturmer, *Journal of Organic Chemistry* 1993, *58*, 5307-5309.

[37] H. Shiraishi, T. Nishitani, S. Sakaguchi, Y. Ishii, *Journal of Organic Chemistry* 1998, *63*, 6234-6238.

[38] A. Gil, L. M. Gandia, S. A. Korili, *Applied Catalysis a-General* 2004, *274*, 229-235.

[39] A. Ekstrom, J. A. Lapszewicz, *Journal of the American Chemical Society* 1988, *110*, 5226-5228.

[40] P. Ciambelli, S. Cimino, S. De Rossi, M. Faticanti, L. Lisi, G. Minelli, I. Pettiti, P. Porta, G. Russo, M. Turco, *Applied Catalysis B-Environmental* 2000, *24*, 243-253.

[41] S. S. Bahekar, D. B. Shinde, *Tetrahedron Letters* 2004, *45*, 7999-8001.

[42] M. Fuming, L. Guangxing, N. Jin, X. Huibi, *Journal of Molecular Catalysis A: Chemical* 2002, *184*, 465-468.

[43] Z. M. Hou, Y. G. Zhang, H. Tezuka, P. Xie, O. Tardif, T. Koizumi, H. Yamazaki, Y. Wakatsuki, *Journal of the American Chemical Society* 2000, *122*, 10533-10543.

[44] L. S. Boffa, B. M. Novak, *Macromolecules* 1994, *27*, 6993-6995.

[45] L. S. Boffa, B. M. Novak, *Tetrahedron* 1997, *53*, 15367-15396.

[46] F. Carree, R. Gil, J. Collin, *Organic Letters* 2005, *7*, 1023-1026.

In: Samarium
Editor: Kaitlyn R. Danford

ISBN: 978-1-63321-045-5
© 2014 Nova Science Publishers, Inc.

Chapter 2

THE SM-COMPLEX/POLYMER COMPOSITE WITH EXCELLENT LUMINESCENT AND X-RAY SHIELDING PROPERTIES

Li Liu, Shipeng Wen and Lu Yao*
State Key Laboratory of Chemical Resource Engineering, Beijing
University of Chemical Technology, Beijing, China
State Key Laboratory of Organic-Inorganic Composites,
Beijing University of Chemical Technology, Beijing, China

ABSTRACT

Sm, a typical member of the lanthanide series, has many functional properties due to its internal particular 4f electronic structure. In particular, the luminescence and X-ray shielding of Sm have attracted considerable interest. In this chapter, we mainly focused on the recent development of Sm-complex/polymer composite because of its excellent functional properties, chemical stability and wide potential application. We also introduce a novel in-situ reaction to prepare the Sm-complex/polymer composite with fine dispersion and strong interfacial interaction. The luminescent composite exhibits a strong yellow color, high luminescent quantum and lifetime. These characteristics enable the composite with a broad application for flexible display, luminescent biosensor and laser systems. The X-ray shielding composite possesses

---

* Corresponding author: Email: LiuL@mail.buct.edu.cn.

less thickness and wide-range energy absorption compared with the classical Pb-contained composite, showing the practical application of human-body protection from X-rays leaking from a nuclear power station, security check, outer space, etc.

1. INTRODUCTION

In recent years, Sm-contained composites are widely applied in the luminescence [1, 2] and X-ray shielding fields [3] due to the unpaired electrons in 4f shell of Sm atom [4, 5]. Research on Sm-complexes has been carried out and the complexes showed unique functional properties such as high luminescent efficiency [6], long lifetime [7] and high shielding ability [8]. Furthermore, many Sm-complex/polymer composites have been prepared and exhibited not only for their excellent functional properties, but also good chemical stability and mechanical performance [9]. He and Tao [10] prepared PVA-Sm composites by the reaction with Sm^{3+} in aqueous solution. The composites, which showed fluorescence peaks around 286nm, were mainly the characteristic luminescence peaks of ligands and strengthened by Sm^{3+} ions. Lin et al. [11] prepared Sm(OPr')(TTA)$_2$/PMMA (Opr=isopropoxide TTA=tris-(2-thenoyltrifluoroacetone)) photoluminescent polymers by bulk coordination polymerization of the Sm-complex and methyl methacrylate (MMA). The composite exhibited strong characteristic luminescence of samarium ion under excitation with radiation of 396 nm.

For obtaining the perfect combination of mechanical and functional properties, it is important to choose an effective fabrication method. The common preparation methods for the rare earth/polymer composite are mainly a simple mixing method and a polymerization method. In the simple mixing method, the rare earths are used as the dispersion phase, but they do not have good compatibility with the polymer matrix. The interface interaction of the composites is poor, and the luminescent property greatly descends as the rare earth concentration increases. The polymerization method includes homopolymerization and copolymerization. The polymerization degree of rare-earth and polymer monomer is always low, due to the effect of the steric clash resulting from the large dimension of the rare-earth complex. Hence, this method cannot tolerate a great deal of rare earths ions in the composite.

Currently, a new method called in-situ reaction has caught more attention [12, 13]. This new method can overcome the poor dispersion and weak interface interaction by blending a low rare-earth complex content in the polymerization method. Generally, the composition in the novel method contains unsaturated rare-earth complex, polymer matrix and peroxide. The in-situ reaction happens at a high temperature used to initiate the polymerization/grafting of the complexes.

Recent studies show that the unsaturated organic Sm-complex particles have a process of "solution-diffusion-polymerization/grafting" in the polymer matrix during the high temperature, leading to a high dispersion degree and strong interface bonding between the fillers and polymer matrix. Hence, the organic rare-earth complex with a very small particle size can be dispersed uniformly in the polymer matrix, and the luminescent efficiency of the rare earth can be improved obviously [14]. This feasible method was further applied, with success, in fabricating novel Sm-complex/polymer composites with excellent luminescent and X-ray shielding properties.

2. A Promising Method to Prepare Sm-Complex/Polymer Composites

As mentioned above, the in-situ reaction process has been reported as the most promising method to prepare unsaturated organic rare-earth complex/elastomer composites with excellent dispersion and comprehensive properties. The following section is the detailed mechanism of the in-situ reaction.

First, an appropriate unsaturated organic rare-earth complex with excellent performance was selected and prepared [15]. Since an unsaturated organic ligand, such as AA, has the in-situ reaction ability because of its C=C double bonds, we introduced them into the polymer matrix (NBR) with peroxide as initiator. The peroxide was dispersed in the macromolecule by blending. The monomer of organic rare-earth complex would start self-polymerization, initiated by peroxide during the vulcanization.

Finally, poly(organic rare earth complex) was gained. But the newly formed poly(organic rare earth complex) had bad miscibility with the polymer matrix, and separated out from the macromolecule matrix. Then, they came together to form poly(organic rare earth complex) particles with the small size. The in-situ method overcomes the former traditional methods' limitation, and can improve the dosage of the rare earth in the polymer matrix.

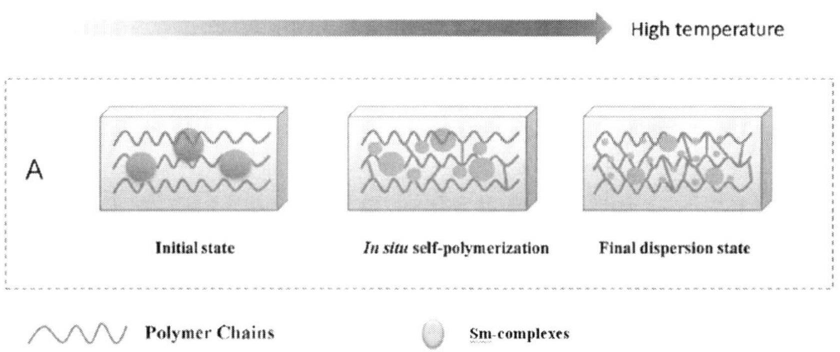

Figure 1. Dispersion mechanism of Sm-complexes in the polymer matrix in situ reaction process.

3. LUMINESCENT PROPERTIES OF SM-COMPLEX/POLYMER COMPOSITES

The rare earth/polymer luminescent materials possess low-cost processing ability, good chemical stability and mechanical strength due to polymer nature. Compared with polymer matrix rare-earth oxide composites, the polymer matrix rare-earth organic complex composites have both excellent processability and luminescent properties. Both saturated and unsaturated Sm-complex/polymer composites have outstanding luminescent intensity, and the unsaturated one has better mechanical properties due to the in-situ reaction process.

3.1. Saturated Sm-Complex/Polymer Luminescent Composites

The saturated Sm-complex/polymer luminescent composite was often prepared using the traditional blending procedure. For elastomer matrix, the complexes were first milled into powders, and then were added into the polymer matrix using a two-roll mixer, several times. Then the curative was added into the blends and the blends were continually mixed until the curatives were well dispersed in the blends. Finally, the blends were vulcanized into film at high temperature according to the vulcanized time. For plastic matrix, the rare earth complex was blended into the matrix at a mixer at a high temperature. After that the blends were mixed in rollers with thin gaps for minutes to get fine dispersion of complexes in the blends. Then the blends were molded into the film at high temperature.

Sm(BA)₃/Polyurethane (PU) composite (BA=benzoic acid) [16] was prepared by melt blending $Sm(BA)_3$ with PU. As the content of Sm^{3+} increasing, the composite luminescent intensity first increases to a maximum value, and then decreases (Figure 2). The fluorescence emission spectrum mainly shows a strong peak at 450nm and a relatively weak peak near 840nm, while $Sm(BA)_3$ cannot emit fluorescent light at the excitation wavelength of 370nm. The reason is that there is a good match between the triplet state of PU and excitation energy orbit of Sm, which makes the energy transition from PU to Sm^{3+} possible. Therefore, it is important to choose an appropriate matrix and ligand, which can be easier to coordinate with rare earth ions.

Sm(TTA)₃phen/PMMA composite [17] was also prepared by a traditional blending process. It was found that the fluorescent intensity of the composite increased with the increase of complex content of the Sm(TTA)₃phen, and there was no fluorescent quenching phenomenon when the complex content was up to 25 wt% (as shown in Figure 3).

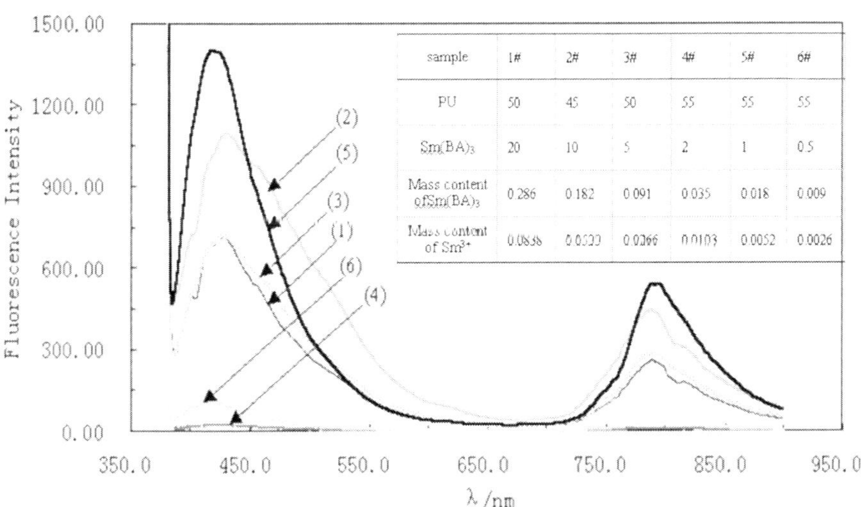

sample	1#	2#	3#	4#	5#	6#
PU	50	45	50	55	55	55
$Sm(BA)_3$	20	10	5	2	1	0.5
Mass content of$Sm(BA)_3$	0.286	0.182	0.091	0.035	0.018	0.009
Mass content of Sm^{3+}	0.0838	0.0533	0.0266	0.0103	0.0052	0.0026

Figure 2. Emission wavelength of Sm(BA)₃/PU(1-6#) (λ_{em}=370nm).

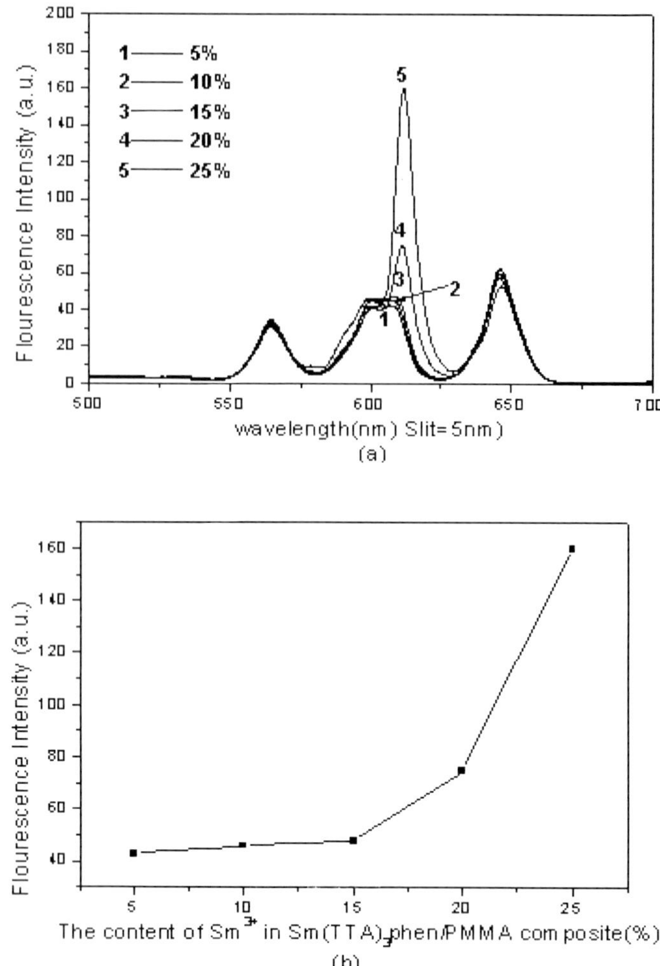

Figure 3. The emission spectra of Sm(TTA)$_3$phen/PMMA composites.

However, the fluorescent intensity of the Sm(TTA)$_3$phen/PMMA composites is much lower than that of the corresponding Sm(TTA)$_3$phen/NBR composites. It can be found that, for Sm(TTA)$_3$phen/NBR and Sm(TTA)$_3$phen/PMMA composites, the multiple emission peaks of the rare earth ions have been depressed and the colorimetric purity of the emission apparently increases. It means that the energy transfer of the organic ligands or the coordination groups of the corresponding polymer matrix becomes more efficient, and the symmetry of the coordinated complex reduces.

For the same rare-earth organic complex, the fluorescence property of the composite is directly related to the dispersion degree of the complexes, interfacial interaction, matrix fluorescent property, and especially the energy transfer from matrix molecules to rare-earth complexes.

NBR has a much stronger interfacial interaction with $Sm(TTA)_3phen$ complexes than PMMA thanks to better compatibility between the $C\equiv N$ group of NBR with $Sm(TTA)_3phen$ than that of the ester group of PMMA. That is one of the key reasons why the $Sm(TTA)_3phen/NBR$ composite has better fluorescent properties than the $Sm(TTA)_3phen/PMMA$ composite.

As to the microstructure aspect, for the NBR-based composite, the complexes disperse uniformly into the matrix. The better coordination between them benefits the energy transfer from the NBR matrix to the $RE(TTA)_3phen$ complexes, preventing the aggregation of the $RE(TTA)_3phen$ particles and therefore reducing the probability of the fluorescent quenching.

3.2. Unsaturated Sm-Complex/Polymer Luminescent Composite

Unsaturated Sm-complex/polymer composite was usually prepared by in-situ reaction process as described before.

$Sm(TTA)_2AA(phen)$/hydrogenated acrylonitrile-butadiene rubber (HNBR) composite [18] was prepared by adding different amounts of $Sm(TTA)_2AA(phen)$ complex and a certain amount of peroxide into HNBR.

Fourier Transform Infrared (FTIR) spectra (Figure 4) showed that the peak intensity assigned to the reactive $C=C$ bond decreased after the curing process, verifying that the in-situ reaction (including polymerization and grafting) of $Sm(TTA)_2AA(phen)$ initiated by the peroxide radical took place during the cross-linking of the HNBR matrix.

Wide-angle X-ray diffraction (WAXD) (Figure 5) testing showed that the crystallinity of $Sm(TTA)_2AA(phen)$ in the composite was dramatically absent after the curing process, implying that most of crystalline $Sm(TTA)_2AA(phen)$ complex took part in the in-situ reaction and formed the non-crystalline $poly(Sm(TTA)_2AA(phen))$. The dispersion phase of cured $Sm(TTA)_2AA$ (phen)/HNBR composites is completely composed of almost nanometer-sized $poly(Sm(TTA)_2AA(phen))$ and few residual $Sm(TTA)_2AA(phen)$ particles with significantly reduced dimensions.

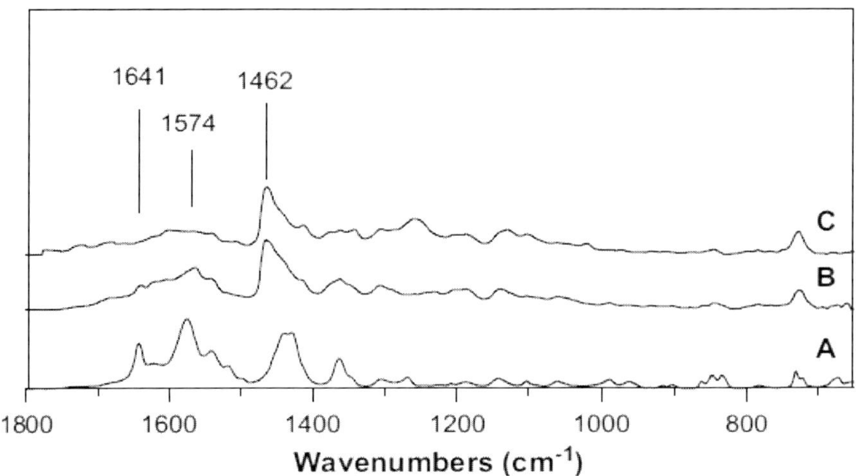

Figure 4. FTIR spectra of (A) Sm(TTA)$_2$AA(phen) complexes, (B) uncured and (C) cured Sm(TTA)$_2$AA(phen)/HNBR composites containing 15 wt % Sm(TTA)$_2$AA (phen) complexes.

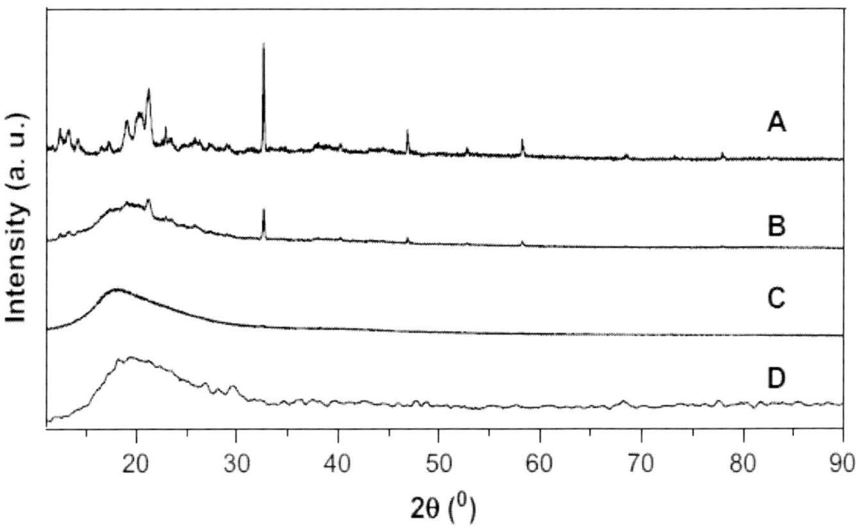

Figure 5. XRD patterns of (A) the original Sm(TTA)$_2$AA(phen). (B) uncured and (cured) cured Sm(TTA)$_2$AA(phen)/HNBR composites containing 15 wt % Sm(TTA)$_2$AA(phen) complexes, (D) cured pure HNBR.

After curing, the size of the dispersed particles in HNBR was further decreased and the largest Sm(TTA)$_2$AA(phen) complex particle size is only about 1 µm (as shown in Figure 5 and 6). The dispersion dimension of the cured composite is far finer than that of the uncured composite. It indicates significantly that the in-situ reaction has a great effect on improving the dispersion of the $Sm(TTA)_2AA(phen)$ complex in the curing process. The particle size of $Sm(TTA)_2AA(phen)$ reduced gradually due to the consumption of substantial $Sm(TTA)_2AA(phen)$ molecules during the in-situ reaction process.

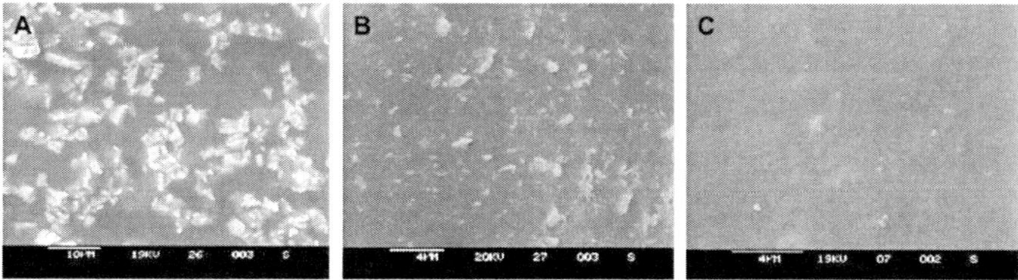

Figure 6. SEM photos of (A) $Sm(TTA)_2AA(phen)$ original particles, (B) uncured and (C) cured $Sm(TTA)_2AA(phen)$/HNBR composites containing 15 wt% $Sm(TTA)_2AA(phen)$.

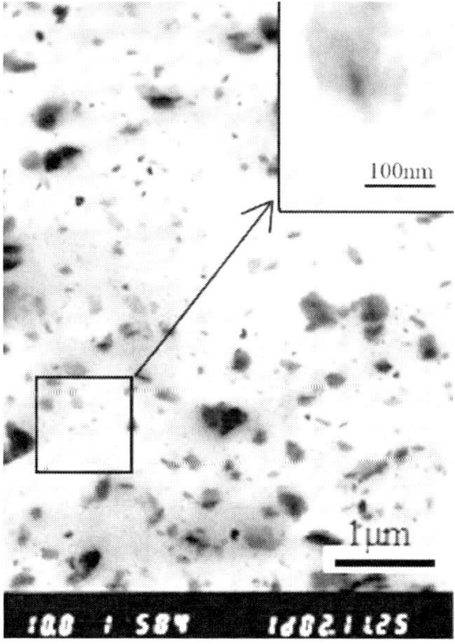

Figure 7. TEM images of the cured $Sm(TTA)_2AA(phen)$/HNBR composites containing 15 wt% $Sm(TTA)_2AA(phen)$ complexes.

Both cured and uncured composites (Figure 7 A and B) possess the similar characteristic emission peaks of the original Sm(TTA)$_2$AA(phen). The fluorescence intensity, around 646 nm of uncured and cured composites, increases with the growth of Sm(TTA)$_2$AA(phen) content. Additionally, from Figure 7C, the fluorescent intensities of cured composites are higher than that of uncured composites. The quantum photoluminescence efficiencies of the cured and uncured samples containing 15wt% Sm(TTA)$_2$AA(phen) measured by the integrating sphere method were 0.7% and 0.3%, respectively. It indicates that in-situ reaction in the cured composites has an effect of improving both the fluorescent intensity and the quantum yields of composites. Therefore, the Sm-complex particles are more finely and uniformly dispersed in the cured composites other than the uncured ones, resulting in a much larger surface area of Sm-complex particles in the cured composites with the same content of Sm^{3+}. Such a fine dispersion level would lead to a higher fraction of Sm^{3+} in dispersed particles that are easily irradiated by the excitation light and emit fluorescence.

For the fluorescence lifetimes of Sm(TTA)$_2$AA(phen)/HNBR (Figure 8), the decay curve of the composites can be fitted by bi-exponential functions, revealing that there are two chemical microenvironments around Sm^{3+} ions [19]. By comparing the decay curve of the uncured and the cured composites, the original chemical microenvironments around the Sm^{3+} ions changed. The site symmetry of Sm^{3+} ions became lower in the cured composite resulting from the formation of the poly(Sm(TTA)$_2$AA(phen)) and the chemical interaction that occurred between the Sm(TTA)$_2$AA(phen) and HNBR matrix during the in-situ reaction. The high asymmetry of chemical micro environments would sensitize the ability of energy transfer of Sm^{3+} ions in the composites and improve the quantum yields of the composites.

Combined in the above analysis, the chemical reaction mechanism during the curing process is shown in Figure 9. First, the peroxide is decomposed to form radicals by heating, inducing the reaction of the double bonds of the Sm(TTA)$_2$AA(phen) monomer and rubber chains. Consequently, the Sm(TTA)$_2$AA(phen) radicals and the rubber radicals are formed respectively. The former initiates its monomer and gives chain growth of poly(Sm(TTA)$_2$AA(phen)) until the chain radicals are terminated by reaction with other radicals. If they are terminated by other poly(Sm(TTA)$_2$AA(phen)) radicals, the separated poly(Sm(TTA)$_2$AA(phen)) molecules are subsequently formed. If they are terminated by rubber chain radicals, the grafted poly (Sm(TTA)$_2$AA(phen)) with rubber chain is obtained. Simultaneously, reactions between rubber chain radicals result in the rubber matrix cross-linking by covalent bonds.

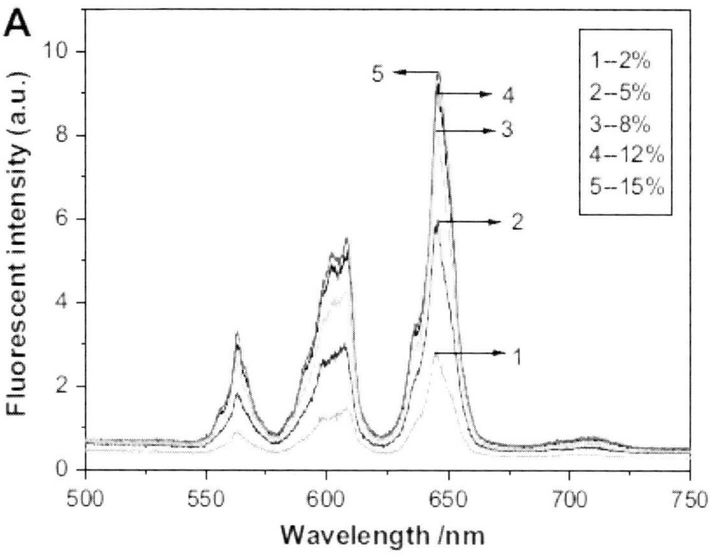

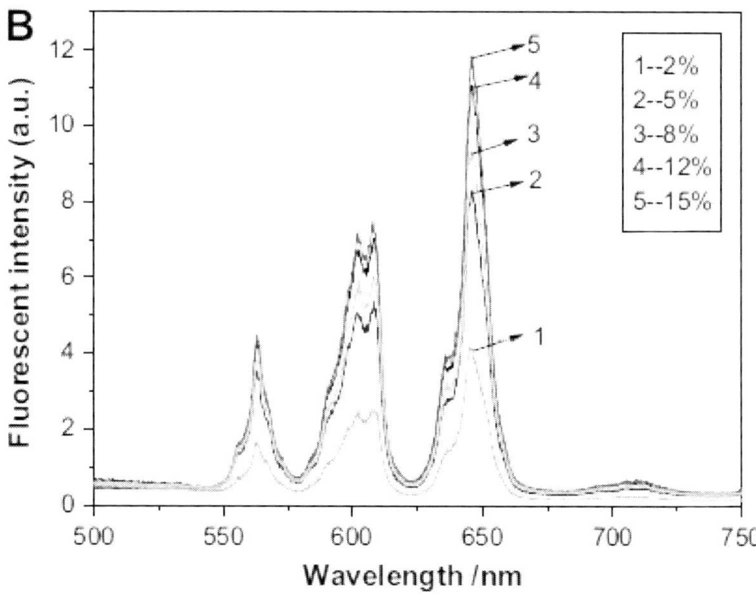

Figure 7. (Continued)

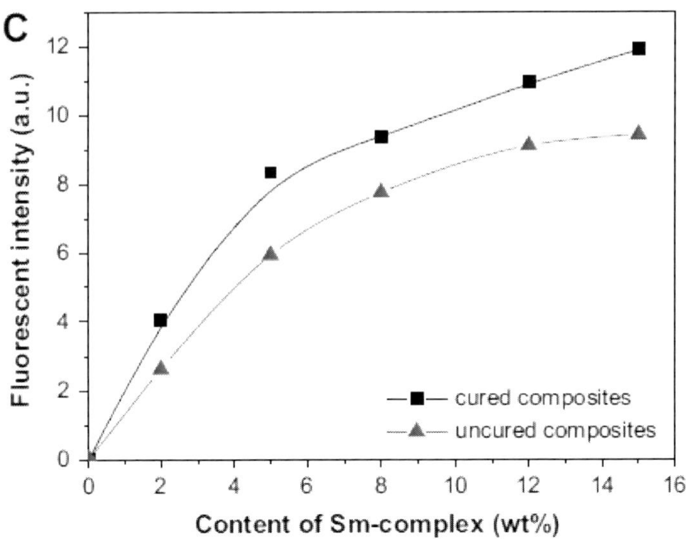

Figure 7. Fluorescent emission spectra (λ_{ex}=385nm) of (A) uncured and (B) cured Sm(TTA)$_2$AA(phen)/HNBR composites with different contents of Sm(TTA)$_2$AA(phen) complexes, (C) a plot of the emission-intensity variation at 645 nm against rare-earth complex content for uncured and cured composites.

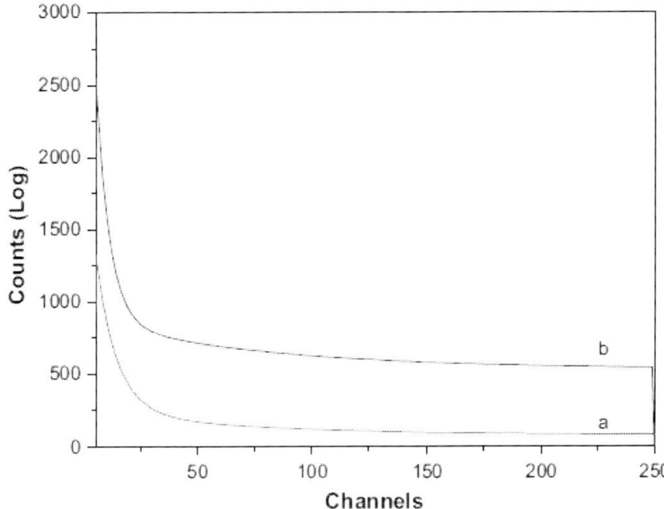

Figure 8. Fluorescent decay curves of (a) uncured composites (Fitted lifetime T1=7.82, T2=1.02) and (b) cured composites (Fitted lifetime T1=17.3, T2=1.37) containing 15 wt % Sm(TTA)$_2$AA(phen) complexes.

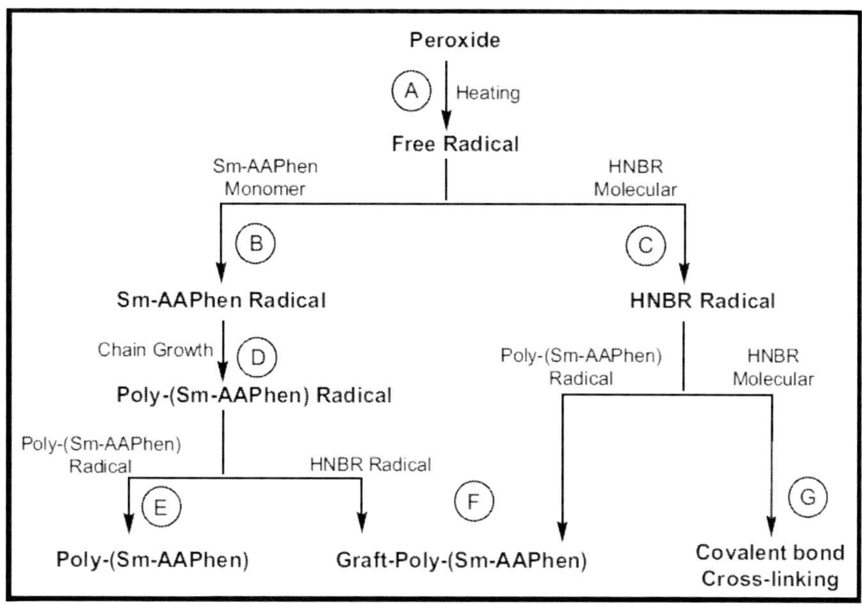

Figure 9. In-situ reaction in the composites during curing: (A) Decomposing of peroxide by heating. (B) Sm(TTA)$_2$AA(phen) is initiated to form Sm(TTA)$_2$AA(phen) radical. (C) HNBR is initiated to form HNBR radical. (D) Chain growth of poly(Sm(TTA)$_2$AA(phen)). (E) Termination of poly(Sm(TTA)$_2$AA(phen)). (F) Graft reaction between HNBR and poly(Sm(TTA)$_2$AA(phen)). (G) Generation of covalent crosslinks.

Sm(AA)$_3$/NBR composite [20] was prepared by two kinds of in-situ reaction. The first in-situ reaction is as described before. In the second kind of in-situ reaction, Sm$_2$O$_3$ and HAA were added to NBR. The mixture was mixed several times in the two-roll mixer. When the compound was mixed equally, the crosslink agent was added in the cold roller to get equal production. Finally, the compound (Sm$_2$O$_3$/HAA/NBR) was cured in the mold at high temperature according to the positive vulcanizing time.

The SEM photograph of the uncured Sm(AA)$_3$/NBR composite (Figure 10a) indicates that the blending of violet mechanical shear stress can help to improve the original particle's dispersion in the high viscosity matrix. In the cured composite, the particle size of Sm(AA)$_3$ reduces obviously, which shows that Sm(AA)$_3$ and the rubber have partly crosslinked with the help of the peroxide in the process of vulcanizing. At the same time, the homopolymerization of Sm(AA)$_3$ monomer has happened, and poly(Sm(AA)$_3$) is gained, shown in Figure 10. Therefore, the in-situ reaction can reduce the particle size of the rare earth complex, which is of benefit to obtain fine dispersion in the Sm(AA)$_3$/NBR composite.

In the Sm$_2$O$_3$/HAA/NBR system (Figure 11 and 12b), the in-situ reaction was carried out more thoroughly, and the particle size of the dispersed phase was much smaller. Therefore, in the course of the blending, in order to make the particle of the rare earth salt disperse more perfectly, the preparation of the composites was very important to the dispersion of the particles.

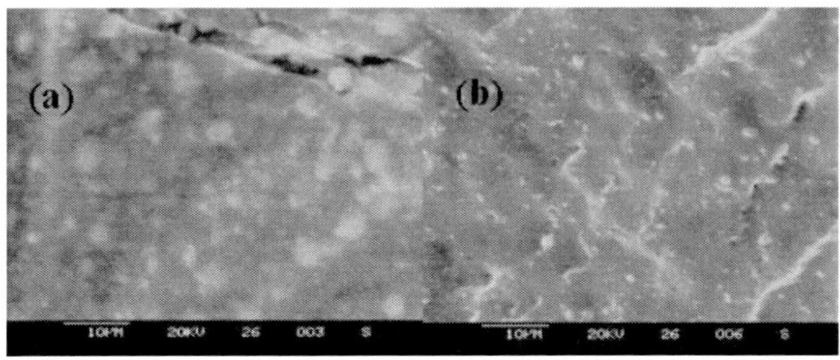

Figure 10. SEM images of (a) uncured and (b) cured Sm(AA)$_3$/NBR composite.

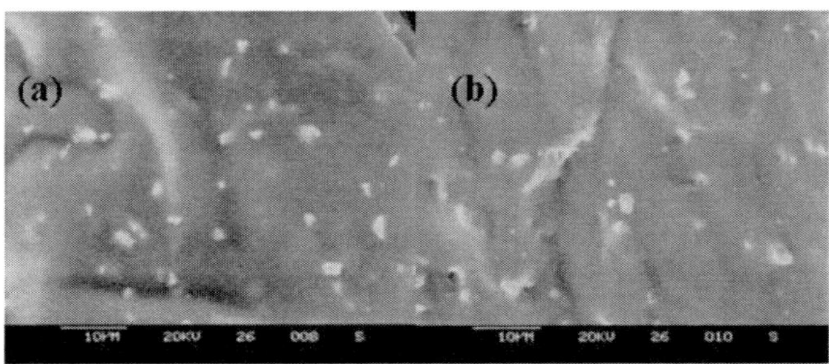

Figure 11. SEM images of (a) uncured and (b) cured Sm$_2$O$_3$/HAA/NBR composite.

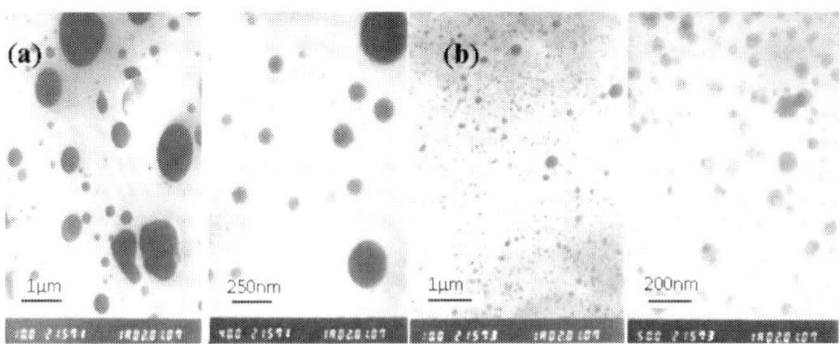

Figure 12. TEM images of cured (a) Sm(AA)$_3$/NBR and (b) Sm$_2$O$_3$/HAA/ NBR composite.

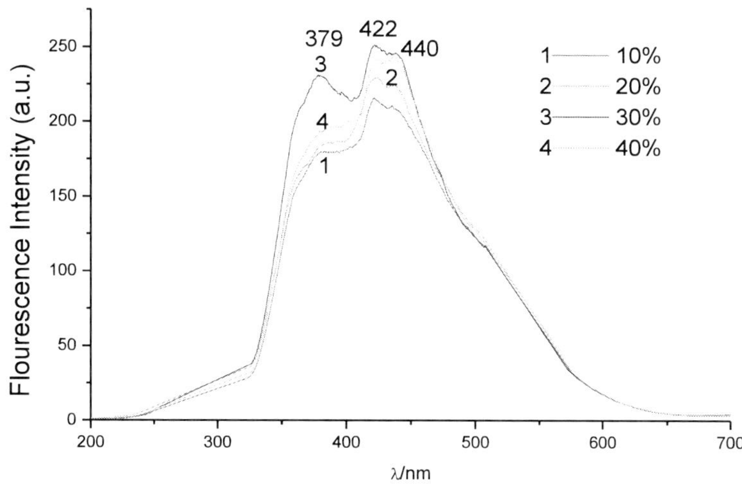

A. Ex=260nm

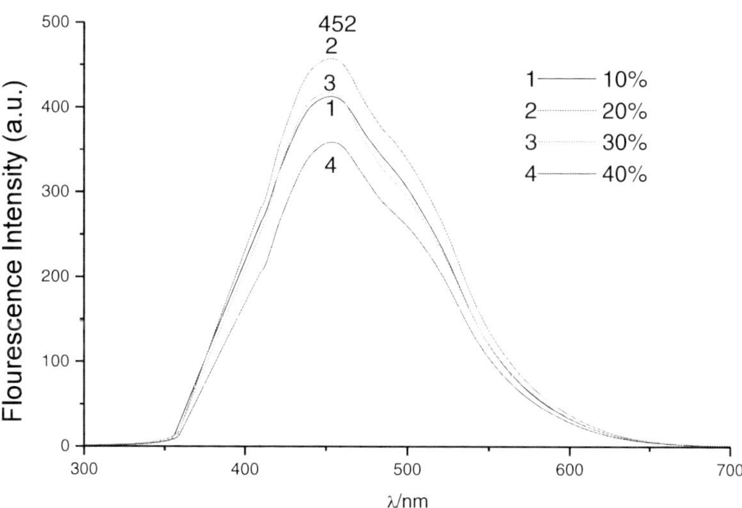

B. Ex=370nm

Figure 13. The Emission spectra of Sm_2O_3/HAA/NBR composite with different Sm_2O_3 contents.

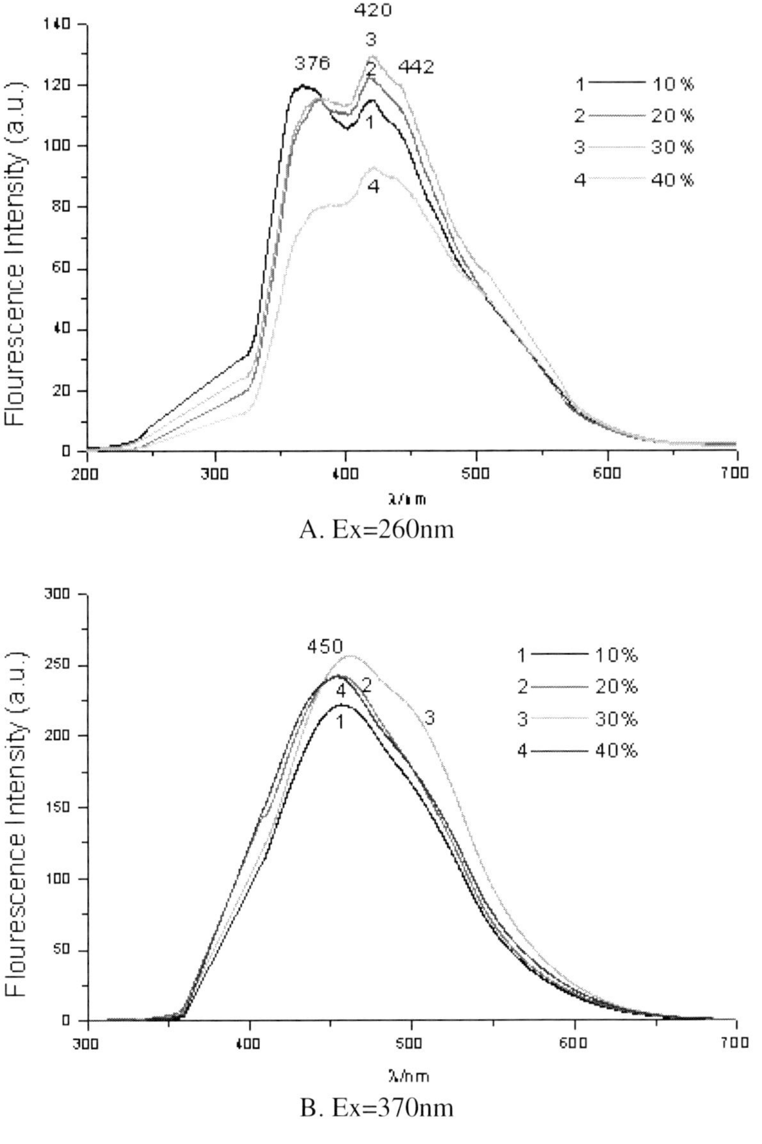

Figure 14. The Emission spectra of $Sm(AA)_3$/NBR composites with different $Sm(AA)_3$ contents.

For the Sm_2O_3/HAA/NBR and $Sm(AA)_3$/NBR composites (Figure 13 and 14), a wide emission peak appears around 450nm. The light energy absorbed by the compound cannot be entirely transmitted to the center Sm^{3+}ion, which emits fluorescence by the form of non-radiation transition. When the exciting wavelength is set at 260nm, the two composites both have two emission apexes, which are shown in Figure13A and Figure14A, respectively. At 379nm and 420nm, the apex and the emission apex of the freedom ion have the same feature. But the characteristic emission apexes at 560nm ($^{4}G_{5/2}{\rightarrow}^{6}H_{5/2}$), 610nm ($^{4}G_{5/2}{\rightarrow}^{6}H_{7/2}$) and 650nm ($^{4}G_{5/2}{\rightarrow}^{6}H_{9/2}$) of Sm^{3+} do not appear. This shows that the triple excited states of the compounds are a bad match with the lowest emission states of Sm^{3+}.

The fluorescent intensity of Sm_2O_3/HAA/NBR composite is much larger than that of $Sm(AA)_3$/NBR composite. The reason is that $Sm(AA)_3$ particles in Sm_2O_3/HAA/NBR composite were mainly formed by the chemical reaction of Sm_2O_3 and HAA during the in-situ reaction. The $Sm(AA)_3$ particle has a much smaller size that that of the $Sm(AA)_3$/NBR composite. The small size of the $Sm(AA)_3$ particle improves the luminescent property. While in the $Sm(AA)_3$/NBR composite, the original big particle of rare earth has a negative effect on the fluorescent property. Therefore, the fluorescent intension is rather small in the $Sm(AA)_3$/NBR composite.

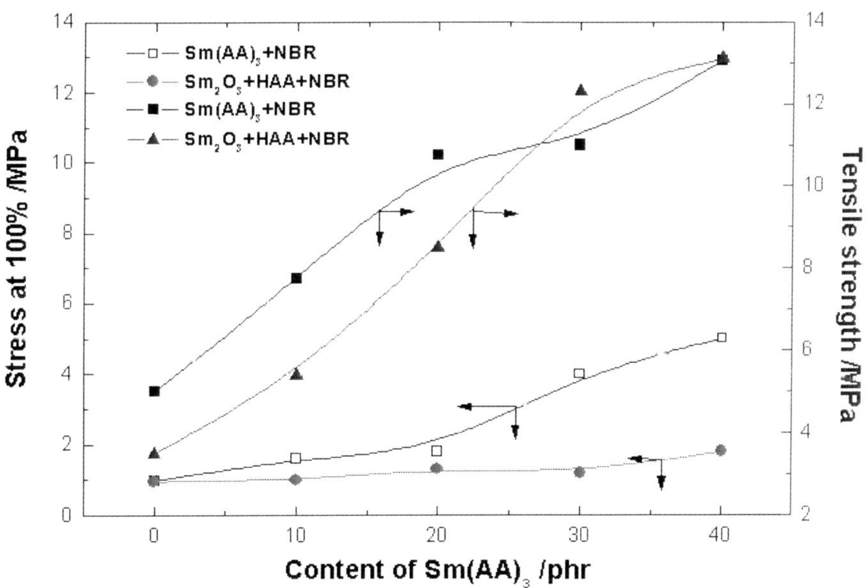

Figure 15. Effect of $Sm(AA)_3$ content on properties of NBR vulcanizations.

From Figure 15, we can see that the mechanical properties of the composite are improved after in-situ reaction. As the amount of HAA and rare earth complex increases, the modules at 100% tensile strength and rigidity raise gradually, but the tear strength changes very little, and the elongation at the break descends, which shows that $Sm(AA)_3$ has a positive effect on enhancing the mechanical property of rubber. Therefore, $Sm(AA)_3$ has a great reinforcing effect on NBR. This could mean that the rare earth complex with double bond can form an additional network with rubber in the course of cross-linking induced by peroxide. At the same time, the rare earth complex can copolymerize and produce nanosized $poly(Sm(AA)_3)$ particles which have an effect on NBR. Hence, the module and tensile strength of the composite are greatly improved.

4. X-RAY SHIELDING PROPERTIES OF SM-COMPLEX/POLYMER COMPOSITE

The ionization radiation is widely used nowadays. It brings great benefits to our modern society but also does some harm to our health, so it is very important to shield the radiation. The essence of ionization radiation is that some microcosmic particles have a certain kinetic energy and can cause ionization. X-ray(γ-ray) has quite a strong penetrating ability. The element Lead (Pb), as a traditional shielding material, has excellent ability to shield radiation with energy higher than 88keV or the energy between 13 to 40keV, but shows poor performance to radiations with energy between 40 to 88keV, which is called "the weak absorbability region". Furthermore, Pb has some other disadvantages such as high density, toxicity, etc.. In order to overcome these shortcomings, lanthanides are selected as radiation-absorption materials substitutes instead of Pb [21]. Li et al. [22] prepared a poly-acrylic acid samarium/epoxy resin composite and found the composite had excellent absorption ability to low-energy and middling-energy radiations. Meanwhile, the composite is non-poisonous and has a lighter density than the Pb-contained composite. Consequently, the polymer-based composite containing lanthanides as radiation-shield materials will gradually replace pure Pb.

Sm_2O_3/thermoplastic polyurethane (TPU) composite [23] was prepared by mechanical blending; the radiation-proof property of the composite was improved with the increase of the amount of Sm_2O_3 under the same testing voltage. The radiation-proof property of the composite at 120 kV was better than that at 60 kV when the amount of Sm_2O_3 is constant. Meanwhile, the radiation-proof property of the composite was superior to Sm_2O_3/NR and PbO/NR (Figure 16).

Sm(AA)₃/NR composite [24] was prepared by in-situ reaction process, in which Sm(AA)₃ with small size dispersed very equably. To enhance the shielding property of the rare earth/rubber composite, the rare earth complex must disperse uniformly in the rubber matrix. From early works, we know Sm(AA)₃ can have an in-situ reaction in the vulcanization process. The reaction could help Sm(AA)₃ disperse well in the polymer matrix. In the preparation, both sulfur and peroxide were used as crosslinking agents in the composite. The first vulcanization process, called pre-vulcanization, occurred at a high temperature at which sulfur decomposed to form a certain degree of crosslinking network. The second vulcanization process, called re-vulcanization process, occurred at higher temperatures at which DCP decomposed to initiate homo-polymerization, graft-polymerization, crosslinking and other reactions. After the two-step vulcanization, the Sm(AA)₃ particle formed an ionomer and became much smaller than the original particle. This will enhance the filler dispersion in the NR matrix (Figure 17 and 18).

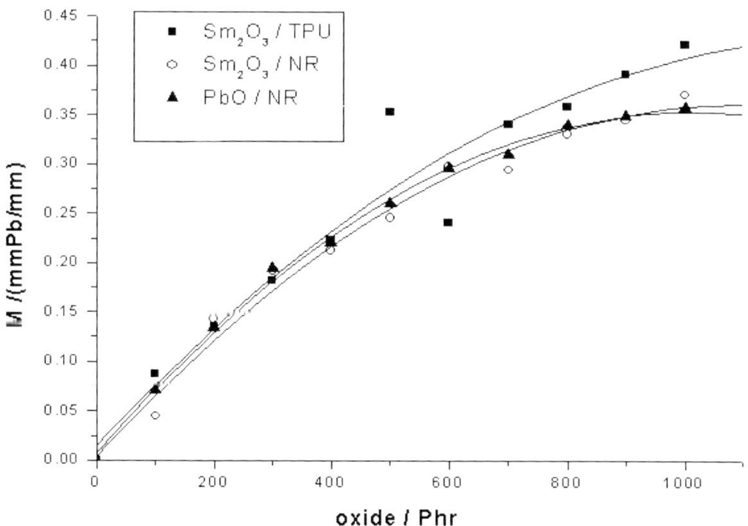

Figure 16. X-ray shielding property of Sm_2O_3/TPU, Sm_2O_3/NR and PbO/NR composites at 120kV.

From Table 1, the composite prepared by two-step vulcanization has a higher value P than that prepared by one-step. It proves the in-situ reaction that happened in the re-vulcanization really has positive effects on $Sm(AA)_3$ dispersion in the NR matrix. Consequently, the radiation shielding property of the composite prepared by two-step vulcanization is also enhanced.

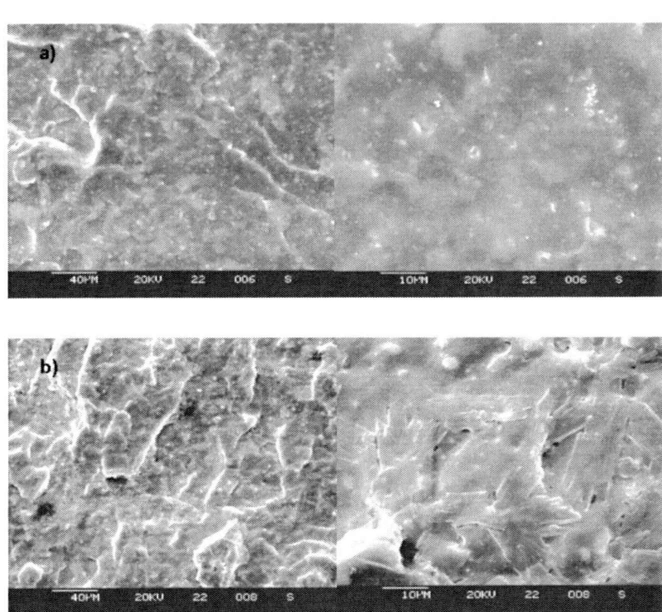

Figure 17. SEM images of (a) uncured and (b) cured $Sm(AA)_3$/NR composite.

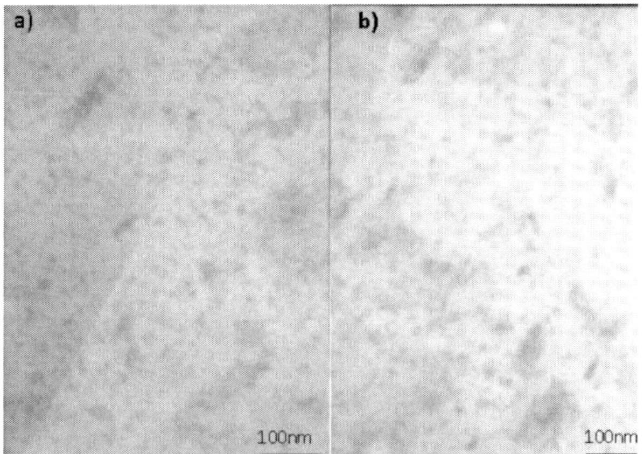

Figure 18. TEM images of (a) uncured and (b) cured $Sm(AA)_3$/NR composite.

Sm(MAA)₃/NBR composite [25] was prepared in the crosslinking stage using sulfur and peroxide, separately. As shown in Figure 19, the tensile property of the composite increased by increasing the Sm(MAA)$_3$content. Furthermore, the tensile strength of the composite, initiated by peroxide, was obviously higher than that of the composite initiated by sulfur (Figure 19). It concluded that the in-situ reaction could hardly be initiated by sulfur, leading to the weak compatibility between the fillers and the polymer matrix and a number of interface defects. While in the composite initiated by peroxide, under the initiation of peroxide free radicals, the C=C bond of Sm(MAA)$_3$ was attacked and further initiated self-polymerization and graft reaction with the rubber matrix. Based on the investigation to the crosslink bond of cured composite, Sm(MAA)$_3$ and the matrix were connected by the ionic crosslinking bond. In consequence, the nano-sized Sm(MAA)$_3$ filler dispersed in rubber.

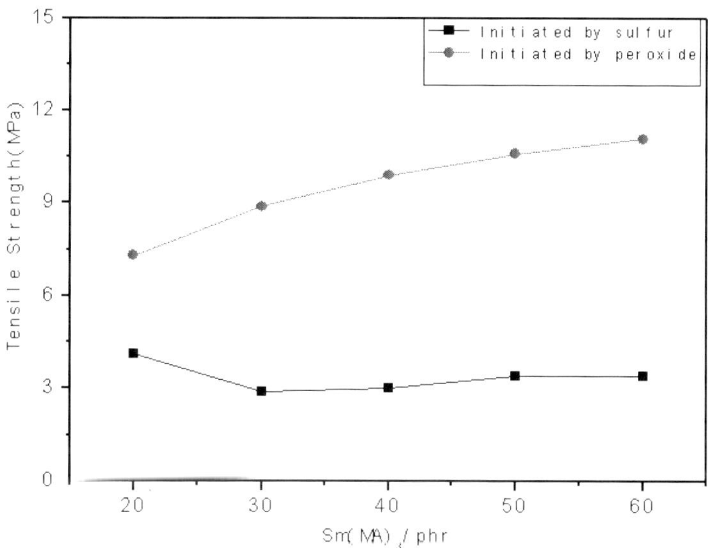

Figure 19. The tensile strength of Sm(MAA)$_3$/NBR composites with different vulcanization system.

Two factors play significant roles in increasing the X-ray shielding property of the composite. First, the high content of effective shielding fillers can lead to excellent X-ray shielding property. Second, fine dispersion of fillers in the matrix would increase the chance of "photoelectric effect" and "Compton effect", which occurs between the shielding material and the incident particles. Therefore, the X-ray shielding property of the composites initiated by peroxide was higher than that of the composites initiated by sulfur (Figure 20).

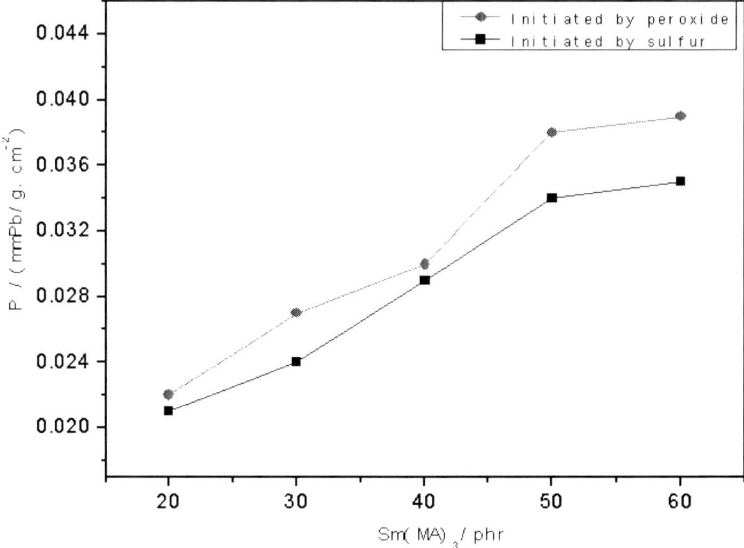

Figure 20. The X-ray radiation shielding property of Sm(MAA)$_3$/NBR composites with different vulcanization system.

Table 1. Radiation shielding property of Sm (AA)$_3$/NR composites

Values	uncured composite	cured composite
$S/$ (mmPb)	0.018	0.020
$\rho/$ (g/cm^3)	1.275	1.305
$h/$ (cm)	0.097	0.098
P/[mmPb/(g/cm^2)]	0.145	0.156

CONCLUSION

Sm-complex/polymer composites combine the outstanding luminescent and/or X-ray shielding property, as well as the fine processing and mechanical performance of the polymer matrix. Moreover, the functional properties and mechanical performance of the composite have both been greatly improved because of the in-situ reaction. These features enable the Sm-complex/polymer composite to have a bright application in the field of imaging materials, lighting devices and X-ray shielding materials.

REFERENCES

[1]   Jimenez, JA; Lysenko, S; Liu, HM; Sendova, M. Luminescence of trivalent samarium ions in silver and tin co-doped a luminophosphate glass. *Opti. Mater.*, 2011, 33, 1215-1220.

[2]   Sakirzanovas, S; Katelnikovas, A; Dutczak, D; Kareiva, A; Justel, T. Concentration influence on temperature-dependent luminescence properties of samarium substituted strontium tetraborate. *J. Lumin.*, 2012, 132.

[3]   Cao, X; Xue, X; Jiang, T; Li, Z; Ding, Y; Li, Y; Yang, H. Mechanical properties of UHMWPE/Sm_2O_3 composite shielding material. *J. Rare Earths*, 2010, 28, 482-484.

[4]   Dorenbos, P; Bos, AJJ; Poolton, NRJ; You, F. Photon controlled electron juggling between lanthanides in compounds. *J. Lumin.*, 2013, 133, 45-50.

[5]   Popov, AI; Plokhov, DI; Zvezdin, AK. Quantum theory of magnetoelectricity in rare-earth multiferroics: Nd, Sm, and Eu ferroborates. *Phys. Rev. B*, 2013, 87, 024413.

[6]   Kaur, G; Verma, RK; Rai, DK; Rai, SB. Plasmon-enhanced luminescence of Sm complex using silver nanoparticles in Polyvinyl Alcohol. *Journal of Luminescence*, 2012, 132, 1683-1687.

[7]   Monteiro, J; Mazali, IO; Sigoli, FA. Determination of Judd-Ofelt Intensity Parameters of Pure Samarium (III) Complexes. *J. Fluor.*, 2011, .21, 2237-2243.

[8]   Ma, T; Liu, YY; Liu, SZ; Liu, ZG. Research Progress of Radiation-proof Material. *Polymer Bulletin*, 2012, 9, 81-86.

[9]   Karah, EK; Daniel, TL; Clare, ER; Paula, MC; Ana, BD; Christopher, LC. Uranyl Sensitization of Samarium(III) Luminescence in a Two-Dimensional Coordination Polymer. *Inorga. Chem.*, 2011, 51, 201-206.

[10] He, SW; Tao TX. Synthesis and luminescence studies of poly(Vinyl Acetate)-Sm Eu coordination compound. *Journal of Anhui Polytechnic University*, 2011, 26, 14-16.

[11] Lin, MJ; Wang, W; Zhang, WG. Sm(Opr^i)(TTA)$_2$/PMMA photoluminescent materials prepared by bulk polymerization. *Acta Polymeria Sinica*, 2006, 8, 1013-1018.

[12] Zhai, YH; Wu, WJ; Zhang, Y; Ren, WT. Enhanced microwave absorbing performance of hydrogenated acrylonitrile-butadiene rubber/multi-walled carbon nanotube composites by in situ prepared rare earth acrylates. *Compos. Sci. Technol*, 2012, 72, 696-701.

[13] Liu, L; Lu, Y; He, L; Wan, Z; Yang, C; Liu, Y; Zhang, LQ; Jin, RG. Novel Europium- Complex/Nitrile- Butadiene Rubber Composites. *Adv. Funct. Mater.*, 2005, 15, 309.

[14] Liu, L; He, L; Yang, C; Zhang, W; Jin, RG; Zhang, LQ. In situ Reaction and Radiation Protection Properties of Gd(AA)$_3$/NR Composites. *Macromol. Rapid Commun.*, 2004, 25, 1197-1202.

[15] Wen, SP; Zhou, Y; Yao, L; Zhang, LQ; Chan, TW; Liang, YR; Liu, L. In situ self-polymerization of unsaturated metal methacrylate and its dispersion mechanism in rubber-based composites. *Thermochi. Acta*, 2013, 571, 15–20.

[16] Liu, L; Zhang, XJ; Zhang, LQ. Preparation and fluorescent properties of Sm(BA)$_3$/PU composite. *China Synthetic Rubber Industry*, 2002, 25, 112.

[17] Zhang, W. Study on the preparation of Rare-earth/ rubber composites and their fluorescent properties. Beijing University of chemical technology, 2011.

[18] Wen, SP; Zhang, XP; Hu, S; Zhang, LQ; Liu, L; Influence of in-situ reaction on luminescent properties of samarium-complex/hydrogenated acrylonitrile-butadiene composites. *Polymer*, 2009, 50, 3269-3274.

[19] Liu, HG; Lee, YI; Park, S; Jang, K; Kim, SS. Photoluminescent behaviors of several kinds of europium ternary complexes doped in PMMA. *J. Lumin.*, 2004, 110, 11–16.

[20] Liu, L; Zhang, XJ; Jin, RG; Yang, C; He, L; Zhang, W; Zhang, LQ. Studies on the Preparation of Sm(AA)$_3$/NBR Composites by in situ Reaction Method and Their Fluorescent Properties. *Chemical Research in Chinese Universities*, 2004, 25, 570-574.

[21] Liu, L; Zhang, LQ; Zhao, SH; Jin, R; Liu, ML. Review on Rare Earth/Polymer Composite. *J. Rare Earths*, 2001, 19, 193-199.

[22] Li, JS; Dai, YD; Zhang, Y; Sun, H; Chang, SQ. Preparation and property of polyacrylic acid samarium/epoxy resin materials for radiation-protection. *Atomic Energy Science and Technology*, 2011, 25, 117-123.

[23] Liu, L; Wu, XF; Fan, L; Zhang, LQ. Radiation-proof and rheological properties of samarium oxide/thermoplastic polyurethane composites. *China Synthetic Rubber Industry*, 2008, 1, 54-57.

[24] Liu, L; He, L; Zhang, W; Yang, C; Liu, YD; Jin, RG; Zhang, LQ. Influence of reaction-induced phase decomposition to dispersion of Samarium acrylic acid in rubber and shielding property of Sm(AA)$_3$/NR composites. *J. Rare Earths*, 2004, 22, 85-90.

[25] Zhou, Y. Study on unsaturated carboxylic acid metal salts' homopolymerization and in-situ preparation of methyl acrylic acid samarium/NBR composites. *Master degree thesis*. Beijing University of chemical technology, 2011.

In: Samarium
Editor: Kaitlyn R. Danford

ISBN: 978-1-63321-045-5
© 2014 Nova Science Publishers, Inc.

Chapter 3

ACTIVATION PROCEDURES OF SmI_2 AND ITS UTILIZATION FOR ORGANIC REACTIONS

Aya Yoshimura and Akiya Ogawa

Department of Applied Chemistry, Graduate School of Engineering,
Osaka Prefecture University, Nakaku, Sakai, Osaka, Japan

ABSTRACT

Samarium diiodide (SmI_2) is one of most important single-electron reducing agents in synthetic organic chemistry and commercially available as a SmI_2-THF solution. Samarium diiodide has the high redox potential and can reduce alkyl iodides, alkyl bromides, aldehydes, ketones, sulfoxides etc. However, the reduction with samarium diiodide itself often requires long period of heating. Therefore, several improved methods for enhancement of the reducing ability of samarium diiodide have been developed. For example, hexamethylphosphoric triamide (HMPA) as a Lewis base is the most effective activating reagent by forming a SmI_2-HMPA complex, $[SmI_2(hmpa)_4]$. Combination of SmI_2 and other metals such as typical metals, transition metals, and rare earth metals is also useful for the reduction of alkyl halides, carbonyl compounds, imines, acetates, and group 14 heteroatom chlorides. The addition of acids, bases, and/or proton sources also works to improve the reducing ability of SmI_2, and sometimes stereoselectivity. Moreover, the photoirradiated reduction system of SmI_2 has been developed as a powerful reduction method without addition of any additives. Currently, these activated SmI_2 systems are very important as synthetic methods in organic chemistry.

INTRODUCTION

Samarium diiodide (SmI_2) has been widely employed as a useful single-electron reducing agent in organic synthesis [1] because of its suitable solubility in organic solvent such as THF, appropriate reducing ability, and homogeneous nature. SmI_2 is air-sensitive, but stable under nitrogen and argon. It was prepared by Matigon and Caze by disproportionation of triiodosamarium at most high temperature (800°C) for the first time in 1906 [2]. However, there was no practical use in organic chemistry until Kagan and co-workers developed a convenient method to prepare the dark blue SmI_2-THF solution (0.1 M) from samarium metal and 1,2-diiodoethane in 1977 (Scheme 1a) [3]. After that, a lot of methods for preparation of SmI_2 have been reported. For example, Imamoto reported the synthetic method of SmI_2 using samarium metal and iodine under refluxing THF (Scheme 1b) [4]. Ishii found that SmI_2 equivalent was easily generated from samarium metal, Me_3SiCl, and NaI in acetonitrile under ambient conditions [5]. Concellón and Flowers reported rapid synthesis of SmI_2 using samarium metal and iodine sources at room temperature by sonication [6]. Moreover, Hilmersson found that SmI_2 was prepared rapidly from samarium metal and iodine in high yields using microwave-assisted heating at 180 °C [7].

(a) Kagan (1977)

$$Sm \ + \ ICH_2CH_2I \ \xrightarrow{\text{THF}} \ SmI_2 \ + \ CH_2=CH_2$$

(b) Imamoto (1987)

$$Sm \ + \ I_2 \ \xrightarrow{\text{THF, reflux}} \ SmI_2$$

Scheme 1. Various methods of preparation of SmI_2.

Recently, Procter investigated the effect of water, oxygen, and peroxide on preparation of SmI_2, and found that they had little influence on the preparation of SmI_2. In addition, Procter has developed a straight forward method for reaching a concentration of SmI_2 sufficient to reduce a range of substrates even when they use low quality samarium metal [8]. Moreover, Srogl developed the facile method for preparation of SmI$_2$ using a continuous-flow system [9]. A solution of iodine source (ICH$_2$CH$_2$I or I$_2$) was driven through a short column filled with samarium powder. After increasing the temperature to 185 °C for about 15 min, the dark blue SmI$_2$ solution was formed at ambient temperature, and the resulting SmI$_2$ solution could be conveniently utilized for several reductions as a reducing reagent.

The redox potential of SmI$_2$ was reported (Sm^{3+}/Sm^{2+}) = -1.55 V in water [10], (SmI$_2^+$-SmI$_2$) = -1.41 V in THF [11] by the use of cyclic voltammetry. It is known that SmI$_2$ itself can reduce several functional groups, such as aldehydes, ketones, alkyl bromides or iodides, α,β-unsaturated carbonyl compounds, epoxides, and sulfoxides [3b]. These reductions were applied to various synthetic reactions, for example, C–C bond formation reactions such as Aldol-type reaction [12], Barbier reaction [3b, 13], Reformatsky-type reaction [3b, 14], pinacol coupling [15], ketyl radical-olefin coupling [16], and medium-size ring formation [17]. Since these reactions proceeded under mild conditions, SmI$_2$ was also available for the synthesis of natural products (Scheme 2) [18]. Recently, SmI$_2$ was applicable to the solid-phase synthesis [19] and fluorous synthesis [20]. Although a SmI$_2$-mediated reductive system has been utilized as described above, the reduction with SmI$_2$ alone often requires long period of heating. Therefore, some additives such as hexamethylphosphoric triamide (HMPA), acid or bases, and water are utilized to enhance the reducing ability of SmI$_2$. In addition, we have revealed that the combination of samarium diiodide and samarium metal exhibits higher reducing ability, compared with those of their single systems [21]. Photoirradiation to SmI$_2$-THF solution also dramatically enhance the reducing ability of SmI$_2$ compared with that of SmI$_2$ in the dark [22]. Herein, we describe various methods of activation of SmI$_2$, that is to say, enhancement of the reducing ability of SmI$_2$, and various organic reactions using activated SmI$_2$ reagents.

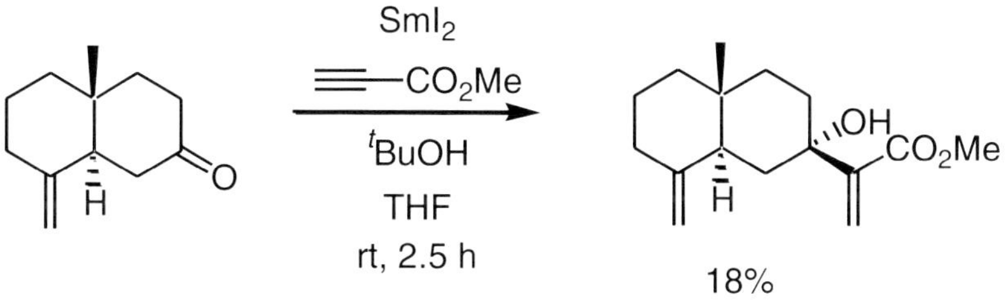

Scheme 2. Synthesis of a natural product with SmI$_2$.

Activation of SmI$_2$ by Lewis Base (SmI$_2$-Lewis Base System)

Hexamethylphosphoric triamide (HMPA) is the most effective activating reagent to enhance the reducing ability of SmI$_2$. Inanaga reported that a remarkable effect of HMPA was observed in the reduction of organic halides with a SmI$_2$-THF solution by the comparison results in the case of SmI$_2$ single system which was reported by Kagan in 1980 (Table 1). In the work of Kagan (without any additives), alkyl chloride could not be reduced despite long-time heating at 67°C. In contrast, in the presence of HMPA, it could be easily reduced under elevated temperature for 8h [23]. In addition, HMPA not only increases the rate of SmI$_2$-mediated reactions, but also enhances the diastereoselectivity of many reactions. For example, in the case of the reduction of 7-octen-2-one, the reductive cyclization proceeded to provide the product, (1R,2R)-1,2-dimethylcyclohexan-1-ol, quantitatively and the diastereomeric exess (de) was 94% in the presence of 8 equivalent of HMPA. In contrast, in the absence of HMPA, (1R,2R)-1,2-dimethylcyclohexan-1-ol was obtained in only 6% yield, and the starting material and byproducts were obtained. The diastereomeric exess (de) was 88% [24].

Table 1. Effect of HMPA on the reduction of alkyl halides

$$R-Cl \xrightarrow[\text{THF}]{\text{SmI}_2} R-H$$

additive	reaction conditions	yield (%)
non	67 °C, 2 days	0
HMPA	60 °C, 8 h	>95

The structural investigation of a SmI$_2$-HMPA complex, [SmI$_2$(hmpa)$_4$], was provided through X-ray structural analysis by Hou, and it was revealed that Sm(II) ion sitted on an inversion center and was bonded to two I$^-$ anions and four HMPA ligands in a distorted octahedron [25]. It was supposed that substrate approach to the samarium metal from either below or above the square plane formed by the HMPA ligands. In addition, the coordination of bulky HMPA to samarium metal led to the stereoselectivity in some reactions. The structure of the [SmI$_2$(hmpa)$_4$] could explain why the use of 4 or 5 equivalents of HMPA to SmI$_2$ was best, and this was confirmed by the reduction of primary carbon radicals [26].

Although the synthetic value of HMPA is excellent, it is known that HMPA has untoward biological effects such as antispermatogenic and mutagenic [27]. Therefore, the reagents alternative to HMPA have been investigated recently. In particular, it was reported that DMPU (*N*,*N'*-dimethyl-*N*,*N'*-propyleneurea) had a low toxicity and was used in some reactions such as the reduction of ketones, alkylhalides, and sulfoxides [24, 28], although the reducing ability of SmI₂-DMPU was lower than SmI₂-HMPA [1m, 29]. The enhancement of the reducing ability of SmI₂ by addition of TMU (1,1,3,3-tetramethylurea) [30], DMI (1,3-dimethyl-2-imidazolidinone) [31], DBU (1,8-diazabicyclo [5.4.0] undec-7-ene) [32], TMG (1,1,3,3-tetramethylguanidine) [32], triethylamine [32], HMDS (hexamethyldisilazane) [33], DPMPA (dipyrrolidinomethyl

aminophosphoric acid triamide) [34], TPPA (tripyrrolidinophosphoric acid triamide) [34, 35] were also reported (Figure 1). In particular, it is suggested that SmI₂ is saturated with four TPPA molecules since the behavior of TPPA was similar to that of HMPA [35a].

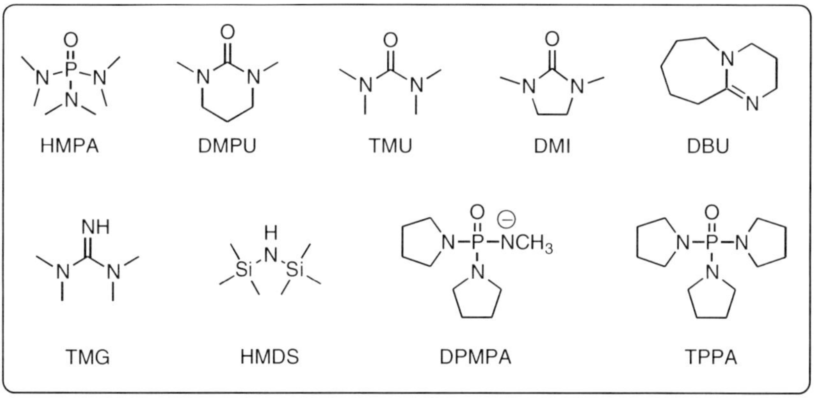

Figure 1. The structures of Lewis base additives.

It was proposed that the reduction using SmI₂ has outer-sphere or inner-sphere electron-transfer process.

The reduction of alkyl halides took place through outer-sphere electron-transfer processes, where as that of ketones took place through inner-sphere processes. Flowers reported that SmI₂-mediated Barbier reactions proceed through outer-sphere electron-transfer processes in the presence of sterically-hindered HMPA. On the other hand, inner-sphere electron-transfer processes were promoted by addition of lithium halide such as lithium bromide (Scheme 3, 4) [36].

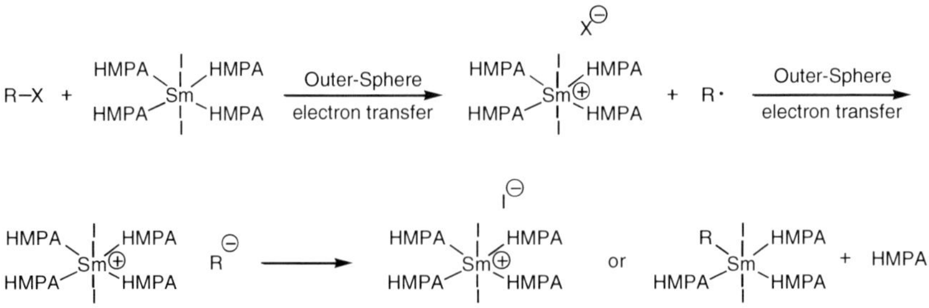

Scheme 3. Outer-sphere electron transfer mechanism.

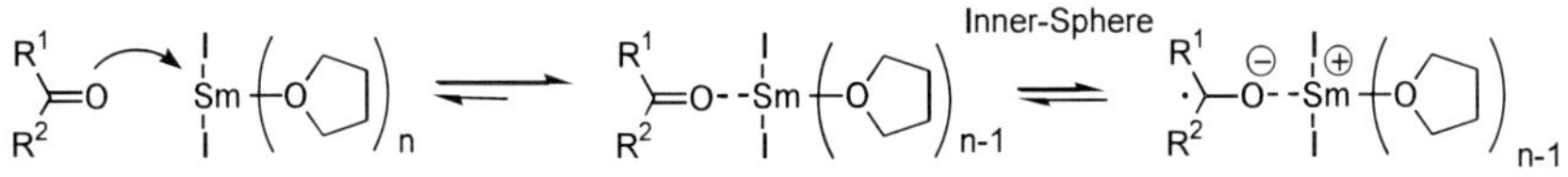

Scheme 4. Inner-sphere electron transfer mechanism.

This suggestion was established from the following results (Table 2). When the SmI_2-mediated coupling reaction of 2-octanone and 1-iodobutane in the presence of HMPA or LiBr was carried out, pinacol coupling product and Barbier product were obtained. The combination of SmI_2 and HMPA gave only the Barbier coupling product. In a sharp contrast, only the pinacol coupling product was obtained by the combination of SmI_2 and LiBr.

Table 2. Coupling data for the reaction of 2-octanone and 1-iodobutane with SmI_2 and different additives

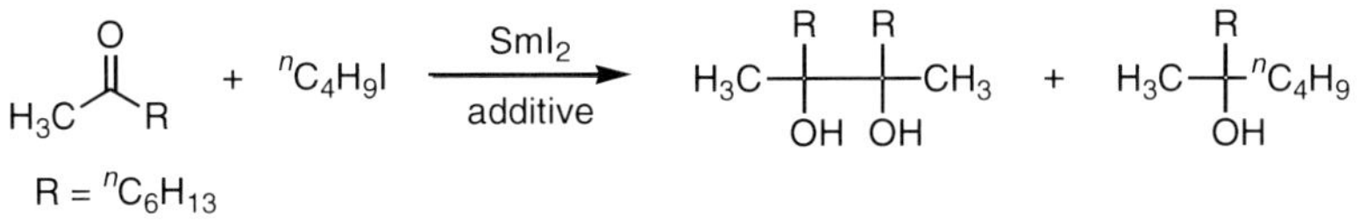

additive	equiv./SmI_2	product formed (%)		unreacted starting material (%)
		pinacol	Barbier	
non	8	23	59	18
HMPA	8	<1	91	8
LiBr	8	98	<1	<1

Activation of SmI$_2$ by Addition of Metals

SmI$_2$-Lanthanide Metals System (SmI$_2$-Ln System)
SmI$_2$-Sm System

We have revealed that the combination of SmI$_2$ and samarium metal enhances the reducing ability of SmI$_2$ compared with those of their single systems [21]. The reduction of bromo- and iodoalkanes with independent SmI$_2$ was reported to require long period of heating at THF reflux (67°C) (Table 1) [3b]. In the presence of both SmI$_2$ and Sm metal, the reduction of 1-bromo- and 1-iodoalkane took place smoothly even at room temperature (Table 3) [21b].

Table 3. Reduction of dodecyl halides with SmI$_2$ and/or samarium metal

$$^nC_{12}H_{25}X \xrightarrow[\text{2-propanol, THF}]{\text{Sm reagent}} {}^nC_{12}H_{26}$$

X	conditions	yield of dodecane (%)		
		SmI$_2$/Sm	SmI$_2$	Sm
I	rt, 5 min.	88	trace	72
Br	rt, 4 h	78	trace	0
Cl	67°C, 20 h	30	10	0

$$R{-}Br \xrightarrow[\text{2-propanol, THF, rt}]{\text{SmI}_2/\text{Sm}} R{-}H$$

Scheme 5. Reduction of alkyl bromides with SmI$_2$/Sm.

In contrast, the reduction of chloroalkanes with SmI$_2$/Sm provided dodecane in low yields. Since SmI$_2$ in the presence of HMPA is reported to reduce chloroalkanes efficiently in refluxing THF (Table 1) [23], the order of reducing ability is estimated tentatively as SmI$_2$/HMPA > SmI$_2$/Sm > SmI$_2$.

The difference of reactivity between bromides and chlorides toward SmI_2/Sm enabled selective reduction of 1-bromo-6-chlorohexane to hexyl chloride. Moreover, SmI_2/Sm is effective for the reduction of secondary and tertiary alkyl bromide (Scheme 5).

We have also found that a reaction system combining SmI_2 with samarium metal accomplished deoxygenative coupling of amide compounds to provide *vic*-diaminoalkenes in excellent yields (Scheme 6) [21a]. Table 4 shows examples of the reductive coupling of various amides. In entry 4, substrate, which includes two amide units, demonstrated intramolecular coupling. Upon coexistence of olefins, three-membered ring was formed; for example, the reaction of *trans*-3-hexenyl-substituted amides gave rise to the three-membered ring product stereospecifically (Scheme 7).

74% ($E/Z = 84/16$)

Scheme 6. Deoxygenative coupling of amide with SmI_2/Sm.

Table 4. SmI_2/Sm mediated reductive coupling of amides

entry	Ar	NR₂	yield (%) (E/Z)
1	Ph	Et₂N	72 (38/62)
2	Cl—	N-piperidine	50 (84/16)
3	naphthyl	N-piperidine	67 (47/53)
4	Ph-C(O)-N~N-C(O)-Ph		62

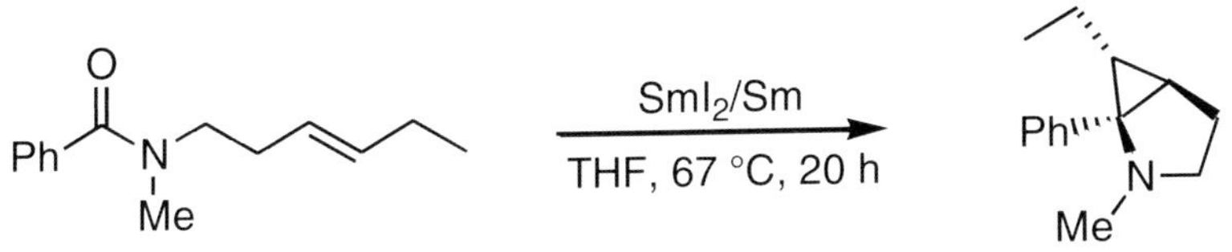

Scheme 7. The formation of three-membered ring.

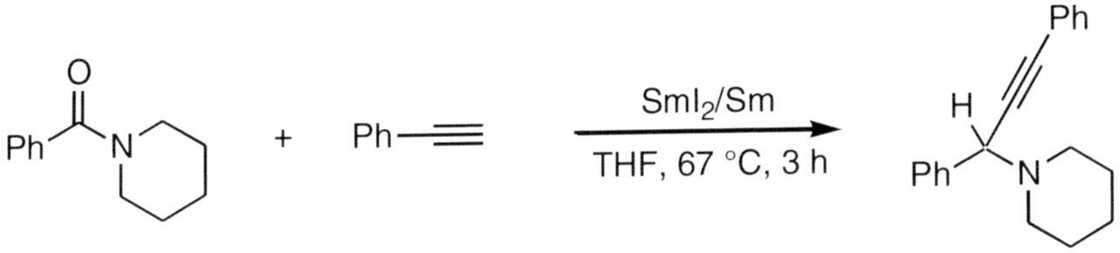

Scheme 8. C-H insertion with benzoylpiperidine and phenylacetylene.

Moreover, the C-H insertion reaction took place upon coexistence of acetylenes; for example, the SmI$_2$/Sm-induced reaction of benzoylpiperidine with phenylacetylene afforded the corresponding propargylic piperidine in good yield (Scheme 8) [21d]. These results led to the suggestion that α-aminocarbene intermediates generated during this reaction.

SmI$_2$-Other Lanthanide Metals System

Namy reported some results on the use of mischmetal in SmI$_2$-catalyzed Barbier-type reactions, reduction of organic halides, pinacol coupling of acetophenone, and the coupling of an acid chloride [37]. Mischmetal is the alloy of the La (33%), Ce (50%), Nd (12%), Pr (4%) and other lanthanides including Sm (1%). It is available at a low price, however, is pyrophoric metal. Scheme 9 shows the results of Barbier-type reactions, in which SmI$_2$ was used in catalytic quantities (0.2 equiv., 10 mol%). Allyl iodide, allyl bromide, benzyl bromide, and ethyl iodide could be reduced easily with SmI$_2$ and mischmetal. A possible reaction pathway was suggested to be through an organosamarium compound Sm(III) species are reduced by mischmetal regenerate samarium diiodide.

Very recently, we investigated the reducing properties of a series of lanthanoid metals in the presence of catalytic amount of SmI$_2$ in the reduction of dodecyl iodide as a model reaction (Table 5) [38]. With scandium (Sc) and yttrium (Y), which are transition metal elements in group 3, reduction of dodecyl iodide with the SmI_2/Ln reagent did not proceed, because they have no 4f-orbital electrons. Lutetium (Lu) also had no reducing activity in the presence of SmI_2. All 4f-orbitals of Lu are fully occupied by electrons, and therefore, Lu is very stable and less reactive. Except for Sc, Y, and Lu, the mixed systems by the combination of other rare earth metals and SmI_2 had intrinsic reducing activities depending on each lanthanoid metal in reduction of dodecyliodide. In a series of reduction systems using SmI_2/Ln reagents, if only SmI_2 acts as a reducing agent and Ln metals solely regenerate SmI_2, the results of reduction would be similar to each other. Very interestingly, however, these results clearly indicate the characteristic features of individual rare earth elements.

Scheme 9. Barbier-type reaction with catalytic SmI_2/mischmetal system.

Table 5. Reduction of dodecyl iodide with SmI_2/Ln reagents

$$^{n}C_{12}H_{25}I \xrightarrow[\text{THF, 2-PrOH, rt, 3 h}]{\text{SmI}_2/\text{Ln reagents}} {}^{n}C_{12}H_{26} + {}^{n}C_{24}H_{50} + {}^{n}C_{10}H_{21}CH=CH_2$$

$$\mathbf{1} \qquad \mathbf{2} \qquad \mathbf{3}$$

	Sc	Y	La	Ce	Pr	Nd	Sm	Eu	Gd	Tb	Dy	Ho	Er	Tm	Yb	Lu
1	0	0	92	93	93	89	86	42	53	57	42	35	22	6	53	0
2	0	0	2	1	1	1	2	0	24	1	0	0	0	0	24	0
3	0	0	2	2	0	0	0	0	3	1	0	0	0	0	3	0

SmI_2-Typical Elements System

The magnesium metal has a reducing power similar to that of samarium metal: (Mg^{2+}/Mg = -2.37 V, Sm^{3+}/Sm= -2.41 V). In Scheme 6, it was shown that not only samarium metal but also magnesium metal effected the coupling of the substrate in the presence of SmI_2 [21a]. Nomura and Endo demonstrated a SmI_2-catalyzed reduction of carbonyl compounds with magnesium metal (Scheme 10) [39]. Moreover, it was found that trivalent samarium salts such as

SmI$_3$, SmBrI$_2$, and SmClI$_2$ underwent the reduction by Mg to provide divalent samarium species (Scheme 11). SmI$_2$ was found to act as a catalyst in the reduction where trivalent samarium formed was smoothly reduced by Mg to regenerate divalent samarium salts. Since their studies, this catalytic SmI$_2$/Mg system has been used by many groups [40].

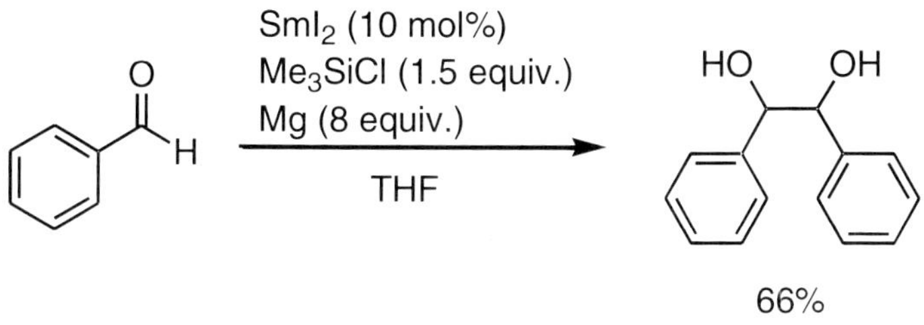

Scheme 10. Pinacol coupling of benzaldehyde with catalytic SmI$_2$ and Mg in the presence of Me$_3$SiCl.

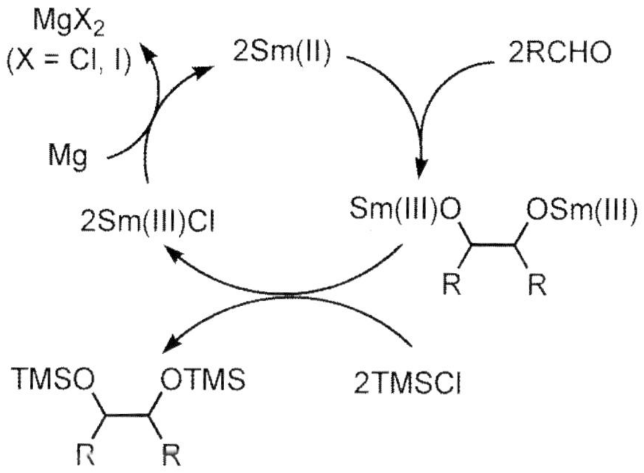

Scheme 11. A possible pathway of catalytic SmI$_2$ and Mg-mediated pinacol coupling.

It was found that amalgamated Al foil, Mg tunings, and Zn powder (20 mesh) reduced effectively SmI$_3$ to SmI$_2$ in THF solution by Corey [41]. The order of the reduction rate was Mg > Al > Zn.

In addition, they developed the SmI$_2$-catalyzed reduction of ketones by the combination of SmI$_2$ (10 mol%), LiI, Me$_3$SiOSO$_2$CF$_3$ (TMSOTf), and Zn-Hg, gave the coupling compounds with acrylates (Scheme 12).

The Reduction of Group 14 Heteroatom Chloride with SmI_2/Sm or Mg System

In view of materials chemistry and organometallic chemistry, application of the C-X bond reduction mediated by SmI_2/metal to E-X (E = Si, Ge, and Sn; X = halogen) bond reduction is of great importance. We found the reduction of group 14 heteroatom chlorine bonds such as Sn-Cl, Ge-Cl, Si-Cl to form Sn-Sn, Ge-Ge, and Si-Si bonds using SmI_2 and Sm and/or Mg metal system [21e, 21f].

Scheme 12. Reaction of ketones with acrylates using SmI_2/Zn-Hg system.

Table 6 shows the results of the reduction of chlorostannane with SmI_2 in the presence or absence samarium metal and magunesium metal. Chloro(tri-*n*-butyl) stannane could be reduced by SmI_2 itself by using excess amounts of SmI_2 (4.0 equiv.) at the THF refluxing temperature (67°C) for 8 h (entry 1); however, the same reduction did not proceed at room temperature for 24 h. In contrast, the reduction of chloro(tri-*n*-butyl) stannane by employing a SmI_2/Sm or Mg-mixed system was examined, which successfully provided the distannane in good yields even at room temperature (entries 2, 3).

Table 6. Reductive dimerization of a chloro(tri-*n*-butyl)stannane

$$^nBu_3SnCl \xrightarrow[\text{THF, rt, 24 h}]{SmI_2/\text{Metal}} {}^nBu_3Sn-Sn^nBu_3$$

entry	SmI_2/metal reagents	temperature	yield (%)
1	SmI_2 (4.0 equiv.)	reflux	76
2	SmI_2/Sm (0.3/0.7 equiv.)	rt	85
3	SmI_2/Mg (0.3/6.0 equiv.)	rt	82

The results of the reduction of chlorotriethylgermane with samarium diiodide and samarium metal or magniesum metal are shown in Table 7.

Table 7. Reductive dimerization of chlorotriethylgermane

$$\text{Et}_3\text{GeCl} \xrightarrow[\text{THF, rt, 24 h}]{\text{SmI}_2/\text{Metal}} \text{Et}_3\text{Ge}-\text{GeEt}_3 \quad + \quad \underset{\overset{|}{\text{Et}}}{\overset{\overset{\text{Et}}{|}}{\text{Et}_3\text{Ge}-\text{Ge}-\text{GeEt}_3}}$$

entry	SmI$_2$/metal reagents	yield (%)	
		digermane	trigermane
1	SmI$_2$ (1.0 equiv.)	No Reaction	
2	Sm (0.7 equiv.)	No Reaction	
3	Mg (6.0 equiv.)	No Reaction	
4	SmI$_2$/Sm (0.3/0.7 equiv.)	73	5
5	SmI$_2$/Mg (0.3/6.0 equiv.)	74	6

As can be seen from entries 1 and 2, an equimolar amount of samarium diiodide could not reduce chlorotriethylgermane at room temperature. In the absence of SmI$_2$, neither samarium nor magnesium metal could work at room temperature for the reductive dimerization of Et$_3$GeCl (entries 2, 3). However, the reductive coupling using a catalytic amount of SmI$_2$ with Sm metal took place efficiently even at room temperature (entry 4). In the case of SmI$_2$/Mg mixed system, digermane was obtained in good yield (74%). In both reactions, amounts of the corresponding trigermane was obtained.

Next, we examined the SmI$_2$/Sm-induced reductive coupling of organochlorosilanes [21f]. However, a ring-opening reaction of THF preferentially took place arising from in situ generation of silyl iodide, which is well-known as a powerful ring-opened silylating reagent for cyclic ethers [21e].

This problem could be solved with use of DME (1,2-methoxyethane) as a solvent, which did not undergo ethereal C-O bond cleavage by silyl iodide under the conditions used for SmI$_2$/Sm-induced reductions. Initially, we chose dimethylphenylchlorosilane as a starting material, and found that the corresponding disilane was obtained in good yield (76%) (Table 8, entry 1). Meanwhile, the reductive coupling of not only aromatic but also aliphatic chlorosilanes was succeeded (Table 8). The reductive polymerization of dichlorosilane was also carried out under the similar reaction conditions (Table 9).

When dichloromethylphenylsilane was treated with SmI$_2$/Sm (0.5 equiv./1.5 equiv.) in refluxing DME for 24 h, the poly(methylphenyl)silane was obtained in 42% yield (entry 1). When magnesium was used as the coreductant, the molecular weight distribution of the poly(methylphenyl)silane (M_w/M_n =1.5, M_n = 4650) was much narrower than that of the polysilane prepared by using a SmI_2/Sm mixed system in the absence of Mg (M_w/M_n =2.6, M_n =1450) (entry 2). However, when samarium powder or samarium diiodide was absent in this reduction system, the reaction failed (entries 3, 4). So far, Wurtz-type coupling of chlorosilanes with alkali metals such as sodium metal has been applied to synthesize polysilanes. However, a Wurtz-type coupling reaction with alkali metals has several serious problems such as difficulty in controlling the molecular weight distribution, as a result of its powerful reducing ability.

Table 8. SmI_2/Sm-induced reductive dimerization of organochlorosilanes

$$R^1-\underset{\underset{R^3}{|}}{\overset{\overset{R^2}{|}}{Si}}-Cl \quad \xrightarrow[\text{DME (4 mL), reflux}]{SmI_2/Sm\ (1.0/3.0\ \text{mmol})} \quad R^1-\underset{\underset{R^3}{|}}{\overset{\overset{R^2}{|}}{Si}}-\underset{\underset{R^3}{|}}{\overset{\overset{R^2}{|}}{Si}}-R^1$$

1.0 mmol

entry	$R^1R^2R^3$SiCl	yield of disilane (%)
1	PhMe$_2$SiCl	76%
2	Ph$_2$MeSiCl	90%
3	Ph$_3$SiCl	80%
4	Ph$_2$(vinyl)SiCl	85%
5	^{t}BuMe$_2$SiCl	81%

Table 9. Reductive polymerization of dichloromethylphenylsilane with SmI_2/Sm system

$$Cl-\underset{\underset{Ph}{|}}{\overset{\overset{Me}{|}}{Si}}-Cl \quad \xrightarrow[\text{DME, reflux, 24 h}]{SmI_2/\text{Metal}} \quad \left(\underset{\underset{Ph}{|}}{\overset{\overset{Me}{|}}{Si}}\right)_n$$

entry	SmI_2/metal reagents	yield (%)	M_w	M_n	M_w/M_n
1	SmI_2/Sm (0.5/1.5 equiv.)	42	3770	1450	2.6
2	SmI_2/Sm/Mg (0.5/0.75/5.0 equiv.)	43	6975	4650	1.5
3	SmI_2/Mg (1.0/20 equiv.)	0	—	—	—
4	Sm/Mg (4/20 equiv.)	0	—	—	—

The method resulted in the formation of the polysilanes with broad and bimodal molecular weight distributions. In contrast, we have succeeded the synthesis of polysilanes with narrow and monomodal molecular weight distribution due to the mild reducing ability of SmI$_2$ reagent. As be seen from the above results, we have developed the reduction of group 14 heteroatom compounds having group 14 element-chlorine bonds, using SmI$_2$/Sm or Mg reagents.

SmI$_2$-Transition Elements System

It was discovered that addition of catalytic Fe(III) salts, for example, FeCl$_3$, Fe(acac)$_3$, and Fe(DBM)$_3$ awfully promoted the various reductions with SmI$_2$ [42]. Alkylation of 2-octanone by butyl iodide took place using SmI$_2$ reduction system, however, it needed 8 h without addition of any additives. In contrast, in the presence of catalytic FeCl$_3$, this reduction proceeded more effectively and 5-methylundecane-5-ol was obtained with the same level of yield (73%) (Table 10). The role of Fe(III) salts might catalyze the reduction of SmI$_3$ to SmI$_2$.

Table 10. Alkylation of 2-octanone by butyl iodide mediated by SmI$_2$ with or without catalytic FeCl$_3$

additive	reaction conditions	yield (%)
non	rt, 8 h	76
FeCl$_3$ (2 mol%)	rt, 3 h	73

Moreover, Kagan screened the catalytic behavior of various transition metal salts and found that NiI$_2$ (1 mol%) was an excellent catalyst for accelerating a lot of reductions by SmI$_2$ in THF such as Barbier reactions, epoxide deoxygenation, pinacol couplings, etc. [43].

This Ni catalytic reduction with SmI$_2$ was expanded to various SmI$_2$-mediated reductions such as a coupling reaction between acid chlorides and esters mediated by SmI$_2$ in the presence of catalytic NiI$_2$, which afforded the corresponding carbonyl compounds were obtained in high yields (Table 11) [44].

Table 11. Coupling between acid chlorides and esters with SmI_2/cat. NiI_2

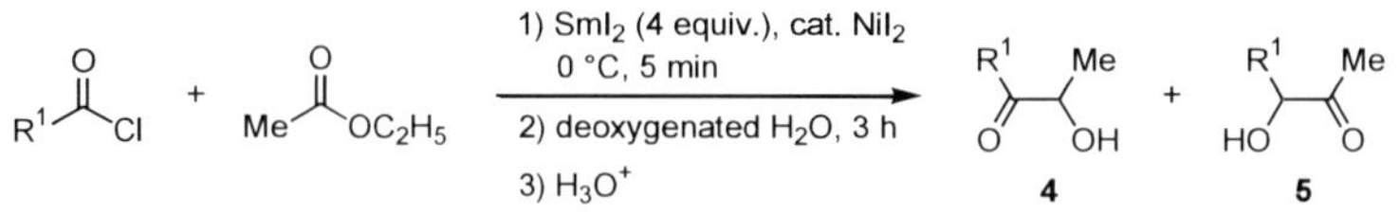

R^1	ester/acid chloride	yield (%)	**4/5**
Ph	10	82	2/92
2-cyclophenyl	10	83	70/30
Cyclohexyl	10	85	33/67

SmI_2/cat. NiI_2 system efficiently promotes homocoupling reaction of imines into vicinal diamines and heterocoupling reactions between imines and ketones (Scheme 13, 14) [45].

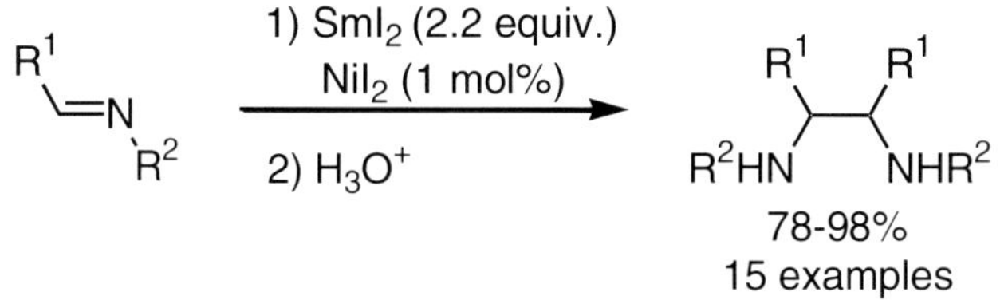

Scheme 13. Reductive coupling of imines to vicinal diamines with SmI_2/cat. NiI_2.

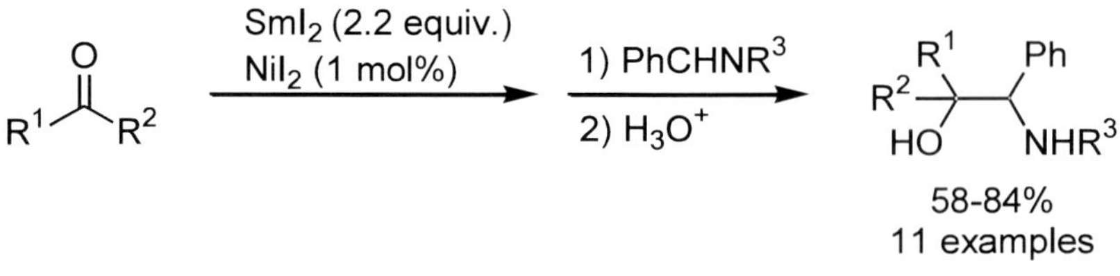

Scheme 14. Reductive coupling of imines and ketones with SmI_2/cat. NiI_2.

This NiI_2 catalyzed reduction with SmI_2 has been also used for the conjugated addition of primary and secondary alkyl halides to α, β-unsaturated esters and amides. The method has been extended to the intramolecular cyclization reactions of alkyl halides with α, β-unsaturated lactone, lactam, and nitrile groups [46]. For example, the halolactone could cyclize to tricyclic lactone in excellent yields with the combination of SmI$_2$ and catalytic NiI$_2$ (Scheme 15).

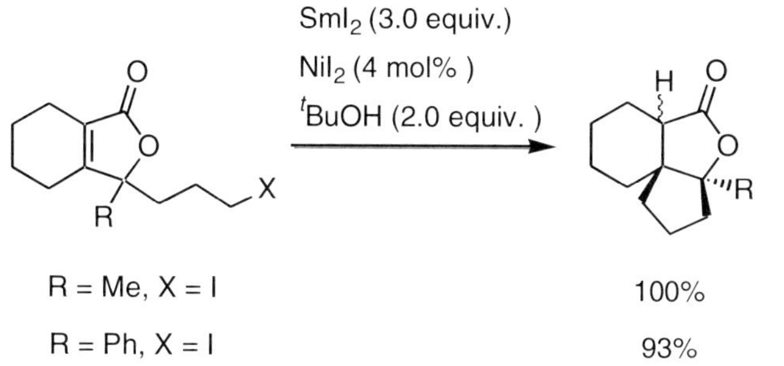

Scheme 15. Intramolecular conjugated additions using SmI$_2$/cat. NiI$_2$ system.

Barbier-type reaction of nitriles and alkyl iodides were mediated by SmI$_2$ in the presence of catalytic NiI$_2$ [47] and this method were applied to synthesize hydroxyl mesylate (Scheme 16) [48].

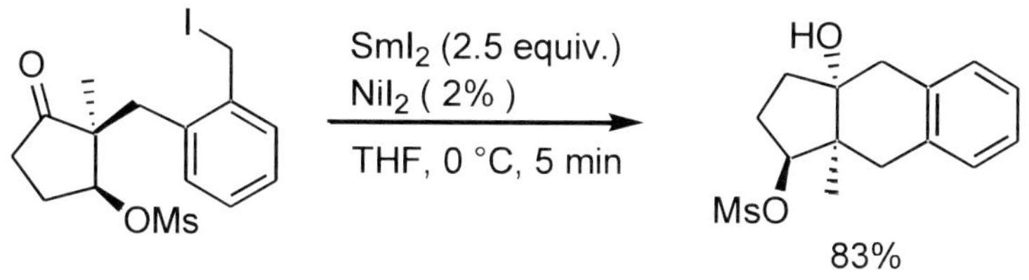

Scheme 16. Synthesis of hydroxyl mesylate with SmI$_2$ and catalytic NiI$_2$.

The reduction systems with SmI$_2$ in the presence of Pd [49], Cu [50], Co [51], and Cr [52] have also been reported. For example, a reductive coupling of allylic acetates with carbonyl compounds and a reduction of propargylic acetates using SmI$_2$ under coexistence of catalytic Pd were reported by Inanaga et al. (Scheme 17, 18) [49a-c].

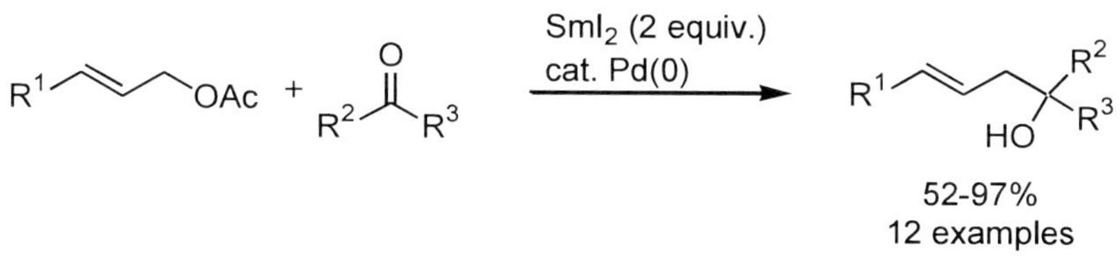

Scheme 17. Reductive coupling of allylic acetates with carbonyl compounds mediated by SmI$_2$ and catalytic Pd.

75-95% (total yields)
11 examples

Scheme 18. Reduction of propargylic acetates with SmI_2 and catalytic Pd.

Activation of SmI_2 by Addition of Bases

It was revealed that carboxylic acids, which are hardly reduced with the ordinary easy-handled reducing agents, were immediately reduced with samarium diiodide in the presence of protic solvents such as MeOH and water by addition of bases at room temperature to the corresponding alcohols in good yields (Table 12) [53]. In the absence of any bases, the reduction of benzoic acid could not proceed (entry 1). In a sharp contrast, the addition of NaOH led to the formation of the corresponding alcohol in an excellent yield (92%) in a very short time (1 min.) (entry 2).

Table 12. Reduction of carboxylic acid with SmI_2 in THF/H_2O/NaOH at room temperature

$$RCOOH \xrightarrow[\text{THF, } H_2O \text{ or MeOH, rt}]{SmI_2 \text{ (4 equiv.), base}} RCH_2OH$$

entry	substrate	base (equiv.)	solvent	time	product	yield (%)
1	PhCOOH	none	MeOH	10 h	PhCH$_2$OH	0.6
2	PhCOOH	NaOH (8)	H$_2$O	60 s	PhCH$_2$OH	92
3	PhCH$_2$COOH	NaOH (8)	H$_2$O	80 s	PhCH$_2$CH$_2$OH	73
4	C$_6$H$_{11}$COOH	NaOH (16)	H$_2$O	60 s	C$_6$H$_{11}$CH$_2$OH	78
5	^{n}BuCH(Et)COOH	NaOH (16)	H$_2$O	10 s	^{n}BuCH(Et)CH$_2$OH	94
6	CH$_3$(CH$_2$)$_4$COOH	NaOH (16)	H$_2$O	5 min	CH$_3$(CH$_2$)$_4$CH$_2$OH	61

As shown in Table 12, not only aromatic carboxylic acids but also aliphatic ones were immediately reduced with SmI$_2$ under the similar conditions to afford the corresponding alcohols in good to high yields. Other bases were investigated and the order of the reduction efficiency was NaOH > LiNH$_2$> NaHCO$_3$> NH$_3$. In addition, this SmI$_2$-base-H$_2$O system can reduce carbonyl compounds, esters, amides, oximes, nitriles, pyridine (Scheme 19), and phenols (Scheme 20, 21) other than carboxylic acids [54].

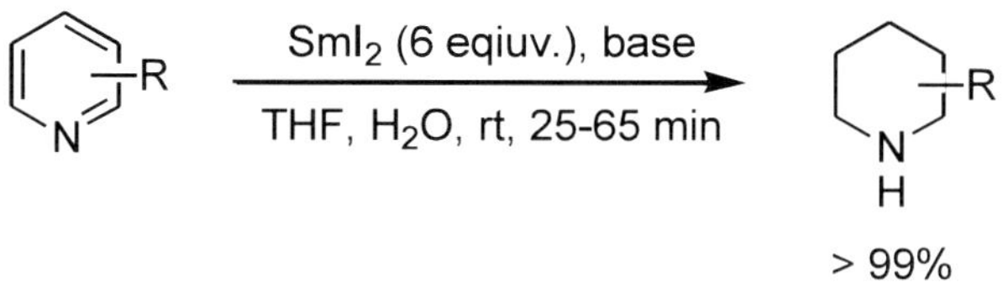

Scheme 19. Reduction of pyridine derivatives with SmI$_2$ and bases.

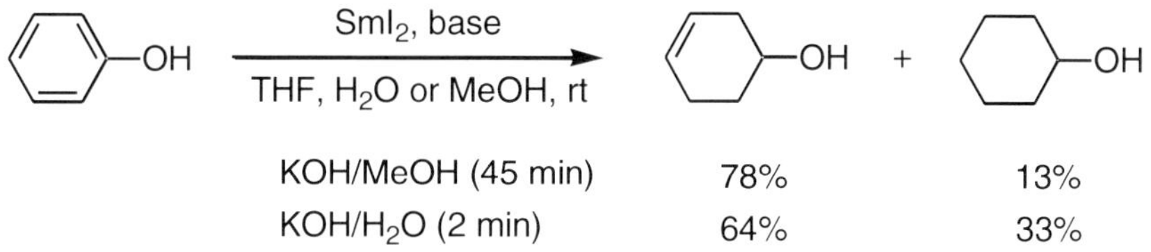

Scheme 20. Reduction of phenol with SmI$_2$ and bases.

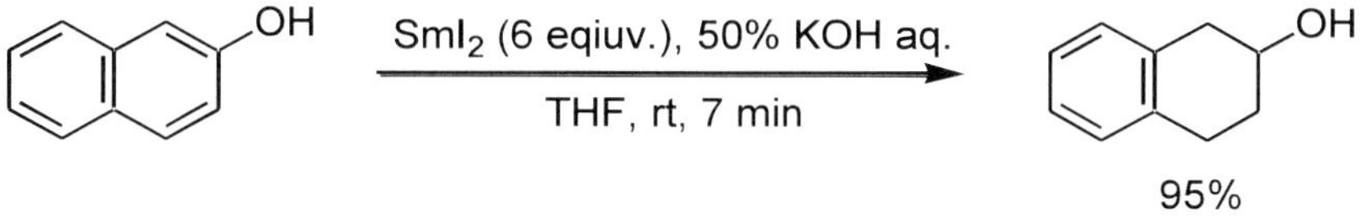

Scheme 21. Reduction of naphthol with SmI$_2$ and bases.

Activation of SmI$_2$ by Addition of Acids

The addition of acids to SmI$_2$ also promoted the reduction of carboxylic acid [55]. In the case of the reduction of benzoic acid, the SmI$_2$-85% H$_3$PO$_4$ system yielded much better results than the other tested SmI$_2$–acid systems as shown in Table 13. Not just only carboxylic acid such as benzoic acid, but esters, acyl chlorides, amides, hydrazides (Scheme 22), and hydroxamic acid (Scheme 23) could also be reduced with SmI$_2$ and 85% H$_3$PO$_4$ to give the corresponding reductants in high yields.

Table 13. Reduction of benzoic acid using SmI_2-acid system

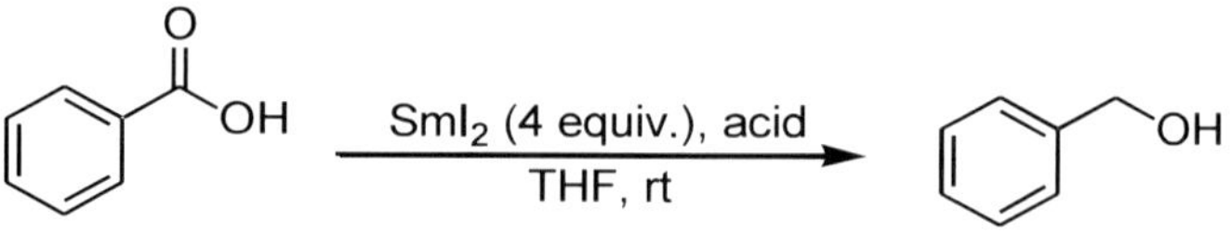

entry	acid	time	yield (%)
1	85% H_3PO_4	3 s	91
2	$H_4P_2O_7$	3 s	68
3	P_2O_5	40 min	3
4	$POCl_3$	3 s	0
5	35% HCl	3 s	75
6	85% H_2SO_4	3 s	48

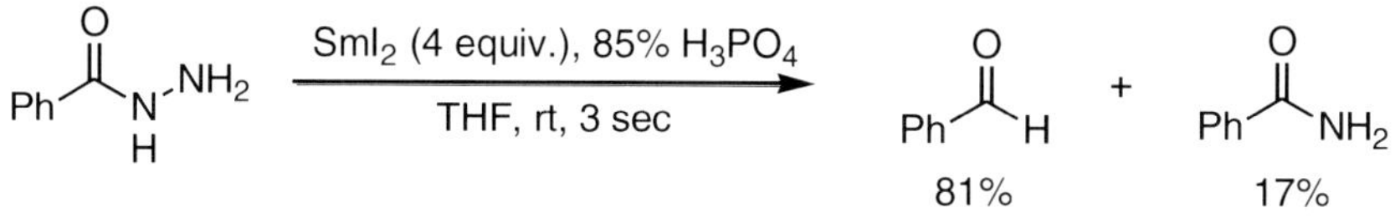

Scheme 22. Reduction of hydrazide using SmI_2-85% H_3PO_4 system.

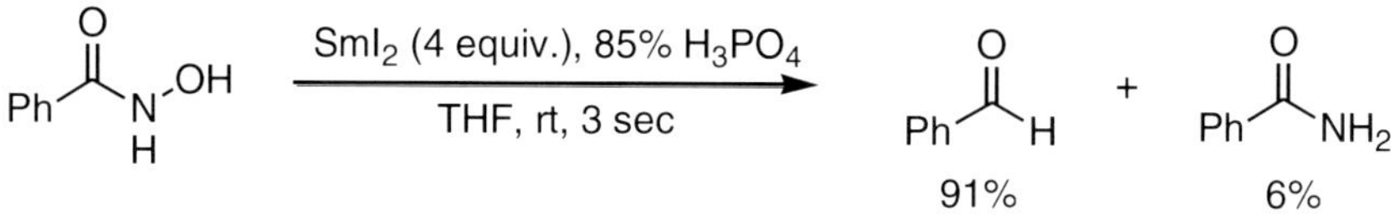

Scheme 23. Reduction of hydroxamic acids using SmI_2-85% H_3PO_4 system.

Methoxy-, chloro-, and alkyl-substituted aromatic dimethyl acetals were reduced with SmI_2 in combination with trifluoroacetic acid (TFA) or $BF_3 \cdot Et_2O$ and vicinal dimethoxyethanes were obtained (Scheme 24). In the cases of cyano- and methoxycarbonyl-substituted aromatic acetals, methyl alkyl ethers were obtained by reductive demethoxylation instead of reductive coupling reaction. It is supposed that the role of TFA or $BF_3 \cdot Et_2O$ could lead to activation of acetals [56]. Suzuki reported that semi-pinacol cyclization of compound, having an acetal and an aldehyde substituent, was achieved with SmI_2 and $BF_3 \cdot Et_2O$. Cyclized product was formed stereoselectivity in high yields (Scheme 25) [57]. In 2011, this SmI_2- $BF_3 \cdot Et_2O$ system was used for reduction of β-hydroxylated pyrrolidine/piperidine aza-hemiacetals and N,S-acetals with α,β-unsaturated compounds in the presence of ^{t}BuOH. This was a key step for synthesis of hyacinthacine A$_2$, uniflorine A (6-*epi*-casuarine), and the unnatural epimer 7-*epi*-casuarine [58].

Scheme 24. Reduction of aromatic dimethyl acetals with SmI$_2$ and, TFA or BF$_3$•Et$_2$O.

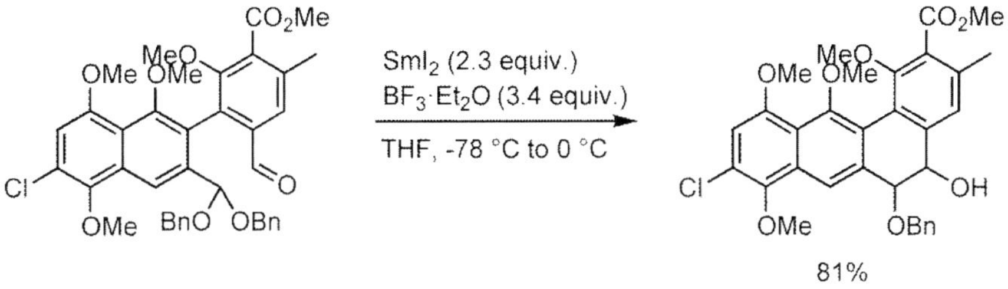

Scheme 25. Reduction of semi-pinacol cyclization with SmI$_2$ and BF$_3$•Et$_2$O.

Furthermore, a convenient method was reported to afford 2,3-dideoxysugar lactons from natural carbohydrates via site-selective deacetoxylation of *o*-acetylsugar lactones by means of SmI$_2$ with pivalic acid or acetic acid (Scheme 26) [59].

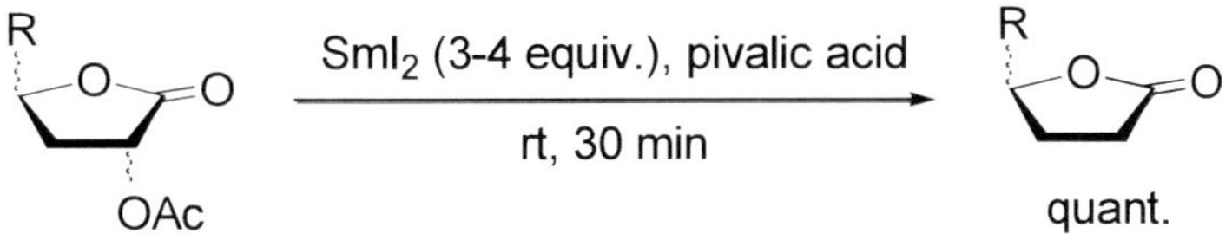

Scheme 26. Deacetoxylation of *O*-acetylsugar lactones with SmI$_2$ and pivalic acid.

Activation of SmI$_2$ by Addition of Water and Alcohol

Many reductions of carbonyl compounds and halides etc. with SmI$_2$ often requires the addition of H$_2$O or alcohols as proton sources. It was found that added water improve the reducing ability of SmI$_2$ by Curran and Hasegawa (Scheme 27) [60]. In the absence of water, 1,3-diphenylacetone was recovered quantitatively after reduction with SmI$_2$. In contrast, corresponding alcohol was obtained in complete conversion in the presence of 33 equivalent of H$_2$O.

Scheme 27. Water effect on the reduction of 1,3-diphenylacetone.

Around the same time, Kamochi and Kudo reported reduction of carboxylic acids, esters, amides, nitriles, pyridine derivatives using SmI_2 in the presence of water [61]. As shown in Table 14, benzoic acid, o-, m-, p-methylbenzoic acids, p-chlorobenzoic acid, and o-aminobenzoic acid were rapidly reduced to corresponding alcohols with SmI_2 (4 equiv.) and H_2O (56 equiv.) at room temperature in excellent yields.

Hilmersson developed the SmI_2-amine-H_2O system for enhancing the reactivity of SmI_2 [1s, 62]. For example, the SmI_2-amine-H_2O mediated reduction of ketones, α, β-unsaturated esters, and imines proceeded instantaneously (Scheme 28) [62a].

This system has been used to reduce a wide range of substrates, for example, alkyl halides [62b], allyl ether protected alcohols [62c], β-hydroxyketones [62d], conjugated olefins [62e], unsaturated hydrocarbons [62f], tosylamides, and esters [62g]. The representative examples of the above-mentioned reductions are shown in Scheme 29. This $SmI_2/Et_3N/H_2O$ system was also applied to selective α-defluorination of polyfluorinated esters and amides (Scheme 30) [62h]. The mechanism of defluorination was suggested in Scheme 31. An initial electron transfer to the π-system of the carbonyl takes place with protonation.

Table 14. Reduction of carboxylic acids using SmI_2 in the presence of water

R	yield (%)
Ph	89
o-, m-, p-MeC$_6$H$_4$	88-90
p-ClC$_6$H$_4$	83
o-NH$_2$C$_6$H$_4$	88

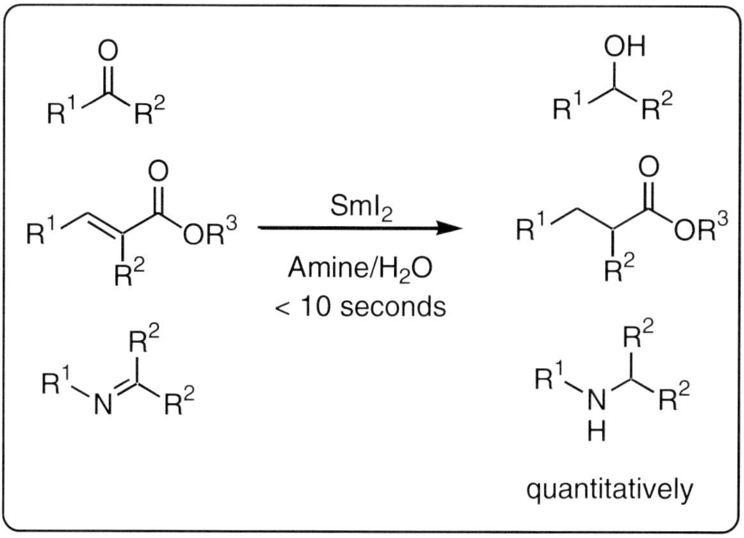

Scheme 28. Instantaneous reduction of ketones, α,β-unsaturated esters, and imines induced by SmI₂-amine-H₂O system.

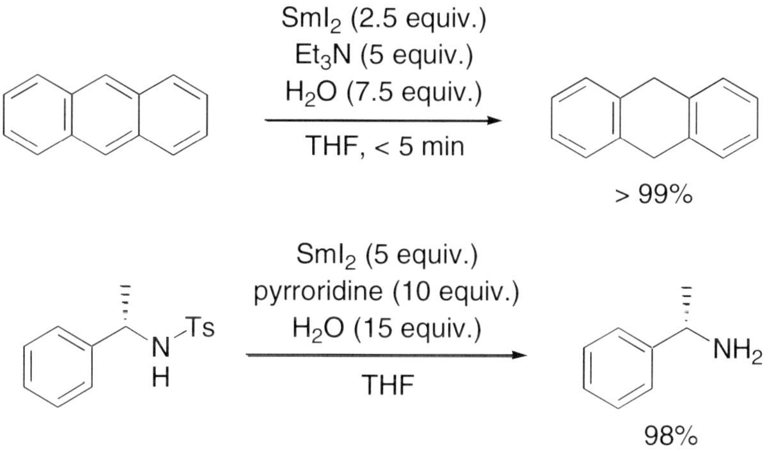

Scheme 29. The reduction of anthracene and tosyl amide with SmI₂/amine/H₂O system.

More recently, Procter et al. reported the reduction of unactivated esters, acids, and nitriles with SmI₂-amine-H₂O system [63]. The SmI₂-Et₃N-H₂O could reduce a wide range of substrates such as primary, secondary, and tertiary alkyl esters and the corresponding alcohols were generated in excellent yields (Scheme 32).

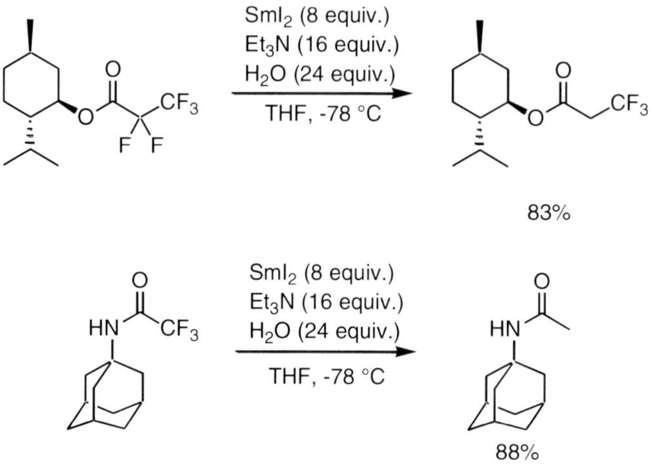

83%

88%

Scheme 30. Selective α-defluorination of trifluoromethyl esters and amides.

Scheme 31. Mechanism proposal for the defluorination.

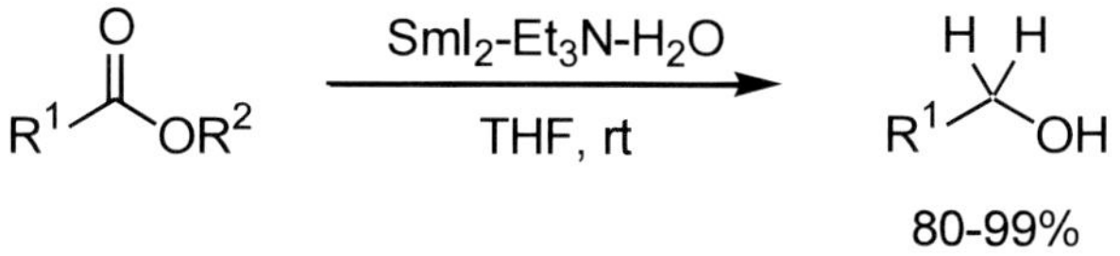

80-99%

Scheme 32. Reduction of unactive esters with SmI_2-H_2O-Et_3N.

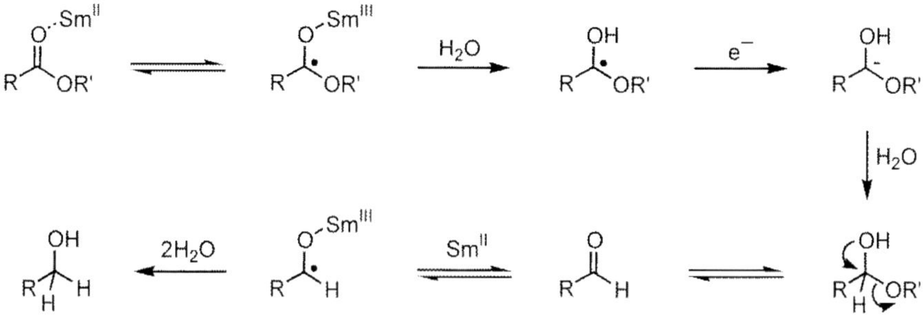

Scheme 33. Proposed mechanism for reduction of esters.

$$71\text{-}99\%$$

Scheme 34. Reduction of unactivated carboxylic acids with SmI$_2$-H$_2$O-Et$_3$N.

$$70\text{-}99\%$$

Scheme 35. Reduction of nitriles with SmI$_2$-H$_2$O-Et$_3$N.

A possible pathway of this reduction is proposed in Scheme 33. Sm(II) coordinates to carbonyl of substrate esters. A radical anion was generated through one electron transfer. After protonation, one more electron transfer and protonation by H$_2$O to generate aldehyde. The aldehyde received one electron from Sm(II), generating the product after protonation. They also reported the reduction of unactivated carboxylic acids and nitriles using SmI$_2$-Et$_3$N-H$_2$O and these reactions are very broad in scope: primary, secondary, and tertiary alkyl carboxylic acids and nitriles were reduced successfully (Scheme 34, 35). Moreover, the reduction of barbituric acids with SmI$_2$-H$_2$O was reported [64].

In 1986, Molander reported the reduction of α-oxygenated ketones and α-heterosubstituted cyclohexane with SmI$_2$ in THF/MeOH solution (Scheme 36, 37) [65].

$$X = OSiMe_3, \ OCOCH_2Ph, \ OTs \qquad 94\text{-}100\%$$

Scheme 36. Reduction of α-oxygenated ketones with SmI$_2$.

Keck et al. screened proton donors such as H$_2$O, MeOH, and ^{t}BuOH in the reduction of β-hydroxy ketones with SmI$_2$ as a model reaction (Table 15) [66a]. Two equivalents of water led to a very fast reaction, which afforded 96% yield of the desired diol as a 83:17 ratio (*anti*: *syn*) (entry 1). In the case of ^{t}BuOH as a proton source, no reaction was observed (entry 2). When 2 equiv. of MeOH was used, the diol was obtained in 95% yield as a 98:2 mixture of *anti* and *syn* isomers (entry 3). Interestingly, the use of excess MeOH (10 eqiuv.) led to a further improved result, i.e., 99% yield of the *anti* product (entry 4). They also succeeded the reduction of β-amino ketones with SmI_2 in the presence of MeOH (Scheme 38) and applied this method to the reduction of β-hydroxy ketone to synthesize epothilones [66b, 66c].

X = Cl, SPh, SO_2Ph, HgCl   76-100%

Scheme 37. Reduction of α-heterosubstituted cyclohexane with SmI_2.

Table 15. Screening proton sources

entry	proton sources	yield (%)	ratio (*anti*:*syn*)
1	H_2O (2 equiv.)	96	83:17
2	^tBuOH (2 equiv.)	No Reaction	—
3	MeOH (2 equiv.)	95	98:2
4	MeOH (10 equiv.)	99	>99:1

Table 16. Initial rates of SmI_2 (0.1 M) mediated reduction of 3-heptanone

entry	proton sources	k (M h^{-1} × 10^{-4})	initial rate rel. to THF
1	THF	0.38	1.0
2	MeOH	1.4	3.5
3	ethylene glycol	16	42
4	di(ethylene glycol)	97	255
5	tri(ethylene glycol)	36	95
6	tetra(ethylene glycol)	2.6	6.9

It was reported that coordination of ethylene glycol to SmI$_2$ promoted the reduction of carbonyl compounds (Table 16). The rate constant of reactions were calculated and the rates were affected by the number of ethereal oxygens in the polydentate alcohols. In particular, di(ethylene glycol) promoted the reduction of 3-heptanone [67]. Mechanistic studies reported by Flowers have shown that initial coordination of di(ethylene glycol) to SmI$_2$ liberates THF or iodide [68]. Procter reported that γ,δ-unsaturated ketones underwent two different stereoselective cyclization with SmI$_2$ depending on the alcohols used (Scheme 39).

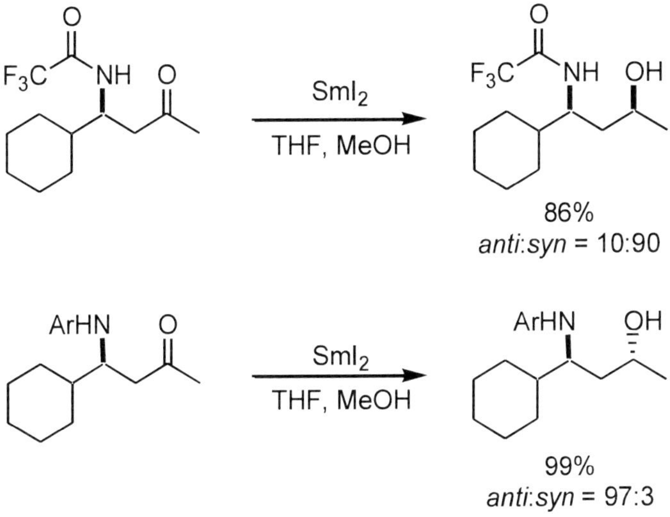

Scheme 38. Reduction of β-aminoketones with SmI$_2$ and MeOH.

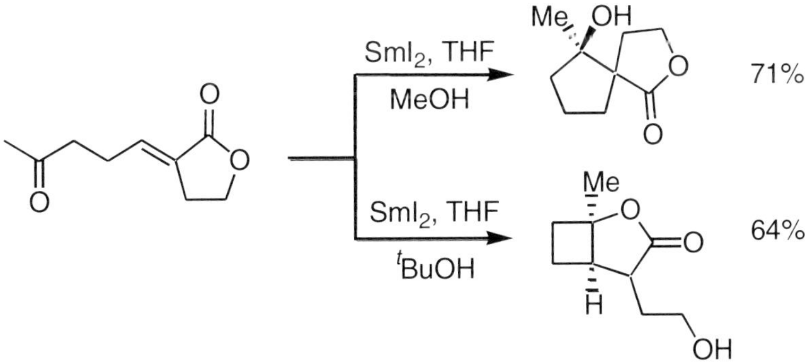

Scheme 39. Stereoselective cyclization of γ,δ-unsaturated ketones with SmI$_2$ and MeOH.

The substrate ketone receives one electron from SmI_2 to generate radical anion. In MeOH, the β-anion is protonated rapidly, and then further single-electron reduction and the following aldol condensation lead to spiro-type product. On the other hand, the rate of protonation to β-anion is slow in the case of ^tBuOH; therefore, the β-anion cyclizes to give cyclobutane derivative via ring-opening and -closure of lacton ring (Scheme 40). Most probably, the reaction may proceed via ketene intermediate 6 [69].

Hoz studied the influence of proton donors on diastereoselectivity and product distribution using the reduction of aromatic and aliphatic ketones [70]. A variety of alcohols such as MeOH, EtOH, trifluoroethanol (TFE), ethylene glycol (EG), and H_2O were used for reduction of norcampher to yield the corresponding alcohol in high *endo* selectivity. When EtOH, TFE, and EG were used as proton donors, the ratio increased with the concentration of the proton donors. In a sharp contrast, H_2O and MeOH yielded a high *endo/exo* ratio at their low concentrations and the ratio decreased as the concentrations of H_2O and MeOH were raised.

6

Scheme 40. A possible mechanistic outline of cyclization of γ, δ-unsaturated ketones.

Activation of SmI_2 by Photoirradiation

The color of SmI_2 in THF was deep blue and the wavelengths of maximum absorption were observed at 565 nm and 617 nm based on transitions from $4f^6$ to $4f^5 5d^1$ [71]. Therefore, we investigated the reduction behavior of SmI_2 upon photoirradiation [22]. Initially, we examined the photoinduced reduction of 1-chlorododecane, which is difficult to be reduced by SmI$_2$ independently, with SmI$_2$ by varying the wavelengths of light to specify the wavelength effecting the higher reducing power of SmI$_2$ (Table 17). While the reduction did not take place upon irradiation with near UV light (300-420 nm) or the light of wavelength greater than 700 nm, irradiation with the light of wavelength between 560 and 800 nm led to the successful reduction of 1-chlorododecane.

These results clearly indicate that SmI$_2$ was activated by irradiating with the light of wavelength between 560 and 700 nm. The procedure can be employed with a variety of organic chlorides such as secondary and tertiary alkyl chlorides and aromatic chlorides, as represented in Table 18 [22a]. These results strongly suggest that the electron transfer rate from photoexcited SmI$_2$ was much faster than that from SmI$_2$ in the ground state. Flowers' mechanistic study indicated excitation enhanced the electron transfer from divalent samarium species [72].

Table 17. Photoinduced reduction of dodecyl chloride

$$^{n}C_{12}H_{25}Cl \xrightarrow[\text{THF, 40 °C}]{\text{SmI}_2\text{-}h\nu} {}^{n}C_{12}H_{26}$$

wavelength	yield (%)
300-420 nm	trace
>500 nm	91
560-800 nm	66
>700 nm	trace
dark	0

Table 18. Photoinduced reduction of alkyl and aryl chlorides

 (2-chlorononane)	88%	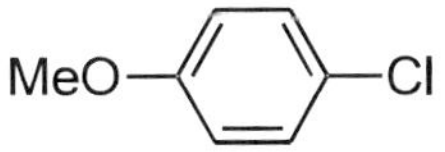 MeO–C$_6$H$_4$–Cl	89%
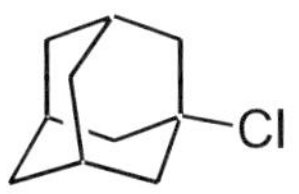 (1-adamantyl chloride)	86%	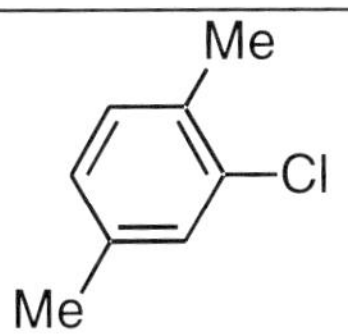 2,4-dimethylphenyl chloride	83%
 PhO(CH$_2$)$_4$Cl	92%	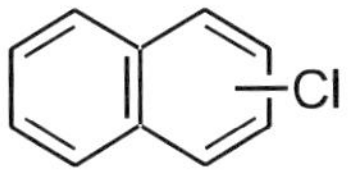 chloronaphthalene	α-Cl: 79% β-Cl: 86%

In the reduction of alkyl chloride, both radical (R•) and organosamarium species ($RSmI_2$) are expected to form as intermediates (Scheme 41). We examined the SmI_2-mediated Barbier reaction of 5-hexenyl chloride (0.1 M) with diethyl ketone upon irradiation with visible light (Scheme 42). The reaction afforded the cyclic and acyclic Barbier adducts with the ratio of 29/71. As the rate constant (k_c) for cyclization of 5-hexenyl radical is 4.19×10^5 s^{-1} (at 40°C) [73], the bimolecular rate constant for the reduction of R• with SmI_2-$h\nu$ is roughly estimated to be 10^5 $M^{-1}s^{-1}$ (at 40°C) [22f]. On the other hand, the first bimolecular rate constant for the reduction of R• was reported to be $0.5\sim7.0 \times 10^6$ $M^{-1}s^{-1}$ (at 40 °C) [74].

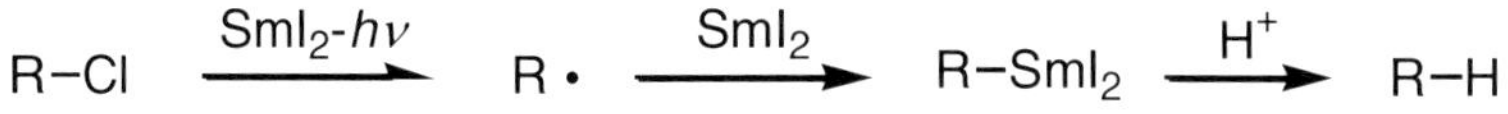

Scheme 41. Photoinduced electron transfer pathway.

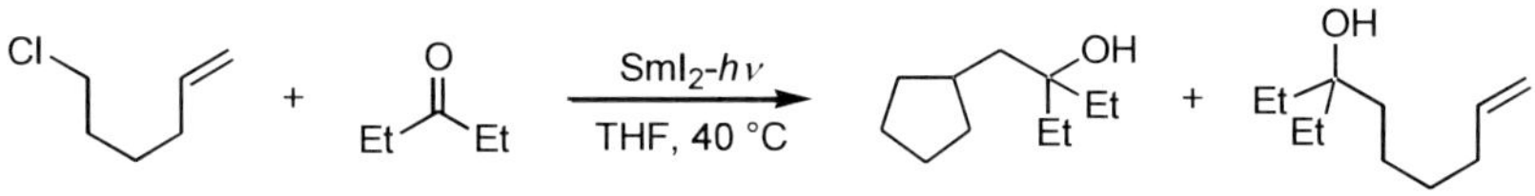

Scheme 42. Barbier reaction of 5-hexenyl chloride upon photoirradiation.

With these kinetic data in mind, we next examined the photoinduced reduction with SmI_2 by the combination of excellent hydrogen donor. Thus, benzenethiol was selected as the hydrogen atom source, because the rate constant for the hydrogen abstraction of primary alkyl radicals from PhSH is determined to be 9.2×10^7 $M^{-1}s^{-1}$ (25 °C) [75]. The photoinduced SmI_2-reduction in the presence of PhSH was successfully demonstrated by the reduction of dichlorocyclopropanes which lead to selective formation of phenylcyclopropane successfully, as indicated in Scheme 43.

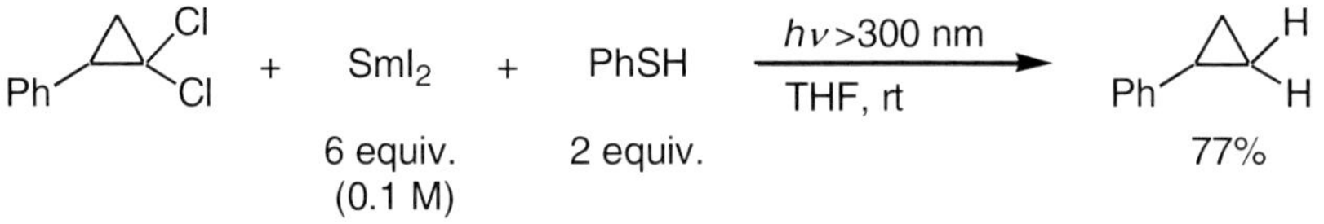

Scheme 43. Reduction of dichlorocyclopropane compound with PhSH.

The SmI_2-$h\nu$ system was found to be useful for the reduction of group 16 heteroatom compounds. The reduction of dodecyl phenyl chalcogenides with

SmI$_2$ was examined upon irradiation with a tungsten lamp through filters and the results were compared with those conducted in the dark [22h, 22i]. As can be seen from Table 19, the SmI$_2$-$h\nu$ system is extremely effective for the reduction of selenides. While the reductive cleavage of C-Se bonds took place exclusively at the C(sp^3)-Se bond, the cleavage of C-Te bonds occurred at both C(sp^3)-Te and C(sp^2)-Te bonds, resulting in lower yield of alkane. The photoinduced reduction of sulfides with SmI$_2$ did not proceed at all under the condition employed (40 °C). Since the starting selenides and tellurides exhibit no absorption in the regions of the wavelength over 400 nm and 500 nm, respectively, the C-Se bond and C-Te bond reduction may proceed *via* electron-transfer from the excited SmI$_2$ to the selenides and tellurides.

Besides dodecyl phenyl selenide, secondary alkyl selenides and heterocyclic selenides underwent reductive deselenation smoothly (Table 20, entries 1, 2). In the case of *cis*-styryl undecyl selenide, the reduction also took place selectively at the C(sp^3)-Se bond (entry 3). Functionalities such as esters and ethers tolerated the reaction conditions (entries 4-6). With bis(3-phenylpropyl) selenide, both 3-phenylpropyl groups could be converted to propylbenzene quantitatively (entry 7). This reaction may proceed *via* the formation of Ph(CH$_2$)$_3$SeSmI$_2$ as a key intermediate. The C-Se bone reduction procedure was also applicable to the reductive cleavage of the C-O bond of organic tosylates, as indicated in entries 8-9. Although the reduction of dodecyl tosylate with SmI$_2$ in the dark afforded only 9% of dodecane, the same reduction upon irradiation with visible light gave raise to 99% of dodecane. Similarly, a tosylated diacetone-D-glucose underwent the reductive cleavage of the C-OTs bond.

Table 19. Reduction of dodecyl phenyl chalcogenides upon photoirradiation

$$^nC_{12}H_{25}X \xrightarrow[\text{THF (20 mL), 40 °C}]{\text{SmI}_2\ (4\ \text{mmol}),\ h\nu} \xrightarrow{H^+} {}^nC_{12}H_{26}$$

X	$h\nu$ (nm)	yield (%)	
		$h\nu$	dark
SPh	>300	No Reaction	—
SePh	>400	99	No Reaction
TePh	>500	38	trace

We also reported carbonylation of organic halides with carbon monoxide mediated by SmI_2 [22a, 22i, 22j]. In general, transition-metal-catalyzed carbonylation of organic halides with carbon monoxide is a synthetically useful and important process in organic synthesis. In contrast, examples of reductive carbonylation of organic halides involving electron-transfer process are still rare.

Table 20. Photoinduced reduction of selenides and tosylates with SmI_2

entry	substrate	product	yield (%)
1	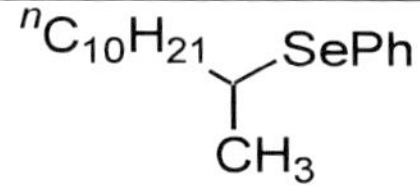	$^nC_{12}H_{26}$	83
2	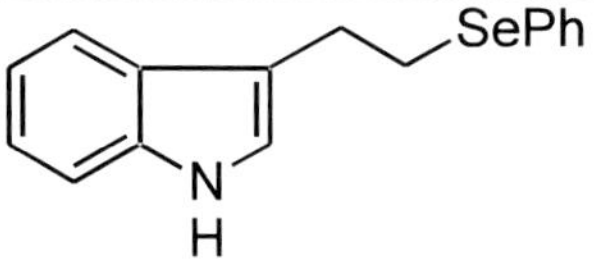	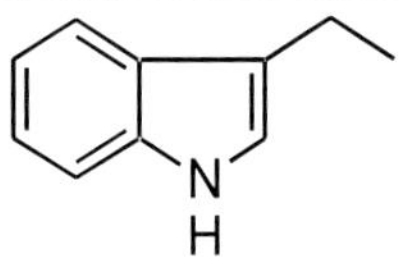	81
3	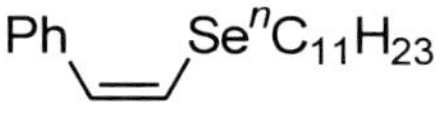	$^nC_{11}H_{24}$	79
4		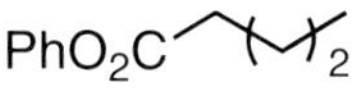	68
5	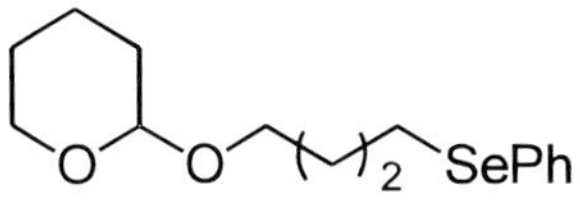	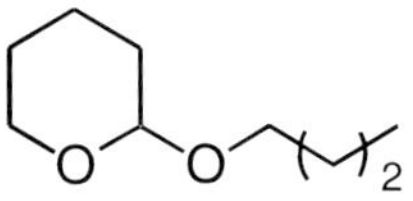	84
6	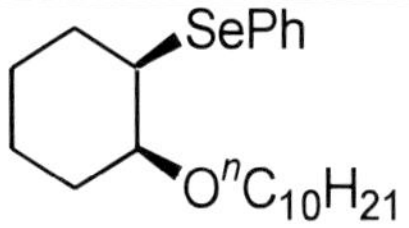	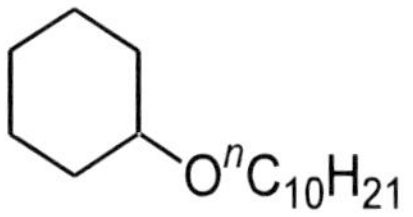	84
7	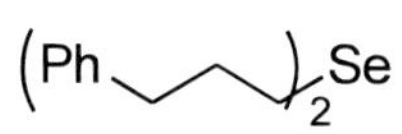	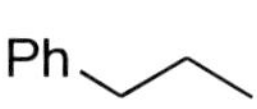	97
8	$^nC_{12}H_{25}OTs$	$^nC_{12}H_{26}$	99
9	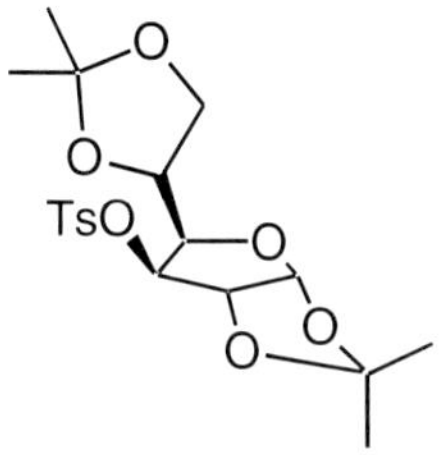	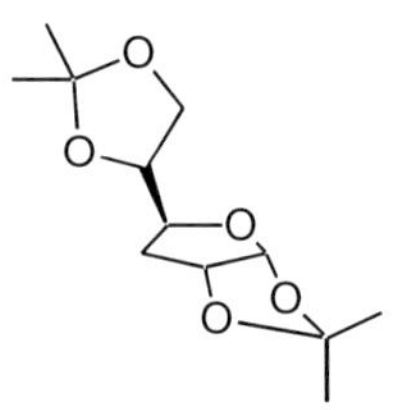	77

Table 21 summarizes the results of the photoinduced carbonylation of several organic halides with SmI$_2$/Sm under an atmosphere of CO. Primary alkyl halides, such as chloro-, bromo- and iodododecanes, produced the corresponding unsymmetrical ketones in moderate yields (entries 1–3). In contrast, no reductive carbonylation took place at all in the dark (entry 4). Cyclohexyl iodide as a secondary alkyl iodide underwent atmospheric carbonylation to afford an unsymmetrical ketone in good yield (entry 5).

Table 21. Atmospheric carbonylation of organic halides

$$R{-}X \xrightarrow[\text{THF (15 mL), rt}]{\text{SmI}_2/\text{Sm (2.0/2.0 mmol), } h\nu, \text{ CO (1 atm)}} R\text{C(O)CH}_2 R$$

0.5 mmol

entry	R-X	yield (%)
1	^{n}C$_{12}$H$_{25}$Cl	45
2	^{n}C$_{12}$H$_{25}$Br	50
3	^{n}C$_{12}$H$_{25}$I	58
4	^{n}C$_{12}$H$_{25}$I (dark)	0
5	^{c}C$_{6}$H$_{11}$I	85

Scheme 44. A possible pathway involving dimerization of acylsamarium species.

The photoinduced carbonylation may involve the formation of acylsamarium species (RC(O)SmI$_2$) as a key intermediate from haloalkane (RX), CO, and SmI$_2$. Kagan et al. previously reported that the reduction of an acyl chloride by an excess amount of SmI$_2$ gave the corresponding unsymmetrical ketones and suggested that a possible reaction pathway might involve the dimerization of acylsamarium species and then further reduction with SmI$_2$ (Scheme 44) [76]. We suggested that dodecyl iodide initially underwent single-electron transfer from SmI_2 to generate a dodecyl radical, which might be captured rapidly by the CO coordinated to the samarium species and simultaneously undergoes single electron reduction by $SmI_2(CO)_n$ generating acylsamarium species as a possible mechanism (Scheme 45).

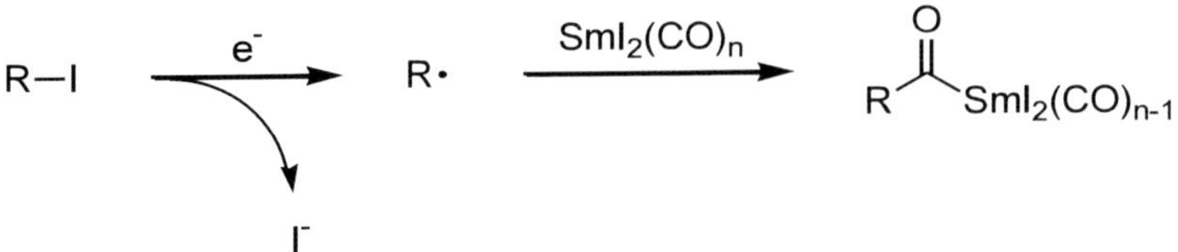

Scheme 45. A possible pathway for the formation of acylsamarium species ($RC(O)SmI_2$).

We have also developed the reductive carboxylation of alkyl halaides with CO_2 by use of photoinduced SmI_2/Sm mixed system [22k]. Although carbon dioxide is a potentially useful carbon resource, practical use in organic synthesis requires appropriate energy because of its energetical stability. Owing to the excellent oxophilicity of SmI_2, CO_2 should coordinate to SmI_2 and undergo reduction upon irradiation with visible light such as sunlight. First, a reductive carbonylation of *n*-dodecyl chloride with SmI_2/Sm upon irradiation with a tungsten lamp was carried out and tridecanoic acid was obtained successfully in 89% yield (Table 22, entry 1). Under similar conditions, various chain-length alkyl bromides and iodides underwent reductive carbonylation, giving the corresponding carboxylic acids in excellent yields (entries 2-5). Secondary alkyl halides such as cyclohexyl halides also gave the corresponding carboxylic acid in moderate yields (entries 6–8). In the case of benzyl chloride, the corresponding carboxylic acid was obtained in 35% yield, despite benzyl radical is a very stable carbon radical species (entry 9). In contrast, tertiary alkyl halides, for example adamantyl iodide, underwent simple reduction to give the corresponding alkane and did not afford the desired carboxylic acid (entry 10). Moreover, the reductive coupling of alkyl fluoride such as *n*-dodecyl fluoride proceeded although the reaction required to elevate temperature (entry 11).

Molander et al. reported a tandem intermolecular carbonyl addition/intramolecular nucleophilic acyl substitution sequence, generating seven- through nine-membered monocyclic, bicyclic, and tricyclic ring systems with good yields and high diastereoselectivities (Scheme 46) [77]. When a mixture of ethyl 2-oxocyclohexaneacetate and 1-chloro-3-iodopropane was treated with SmI_2 and cat. NiI_2, the intermediate chloroalkyl-substituted lactone was smoothly formed within minutes in 91% yield. The second step of the sequential process did not take place directly under these reaction conditions. Addition of an excess amounts of HMPA to the reaction mixture provided the desired bicyclic hemiketal with excellent diasteroselectivity, although the yields were low (36%) with recovery of large amounts of the intermediate lactone. On the other hand, irradiation of the lactone with visible light in the presence of SmI$_2$ promoted the reduction of the chloride to provide the desired hemiketal in 70% yield.

They also reported the cyclization and intramolecular ketone-nitrile coupling took place by photoirradiation [78]. A nucleophilic substitution to form cyclopentanone derivative, followed by the ketyl-nitrile coupling reaction involving 5-*exo* cyclization of ketyl radical species with the nitrile afforded cyclic ketones in 49% yield after hydrolysis (Scheme 47).

Table 22. Visible-light irradiated reductive carboxylation of alkyl halides with CO$_2$

$$R{-}X \xrightarrow[\text{THF (15 mL), rt}]{\text{SmI}_2/\text{Sm (1.5/0.5 mmol), } h\nu, \text{ CO}_2 \text{ (1 atm)}} R{-}COOH$$

0.5 mmol

entry	substrate	product	time (h)	yield (%)
1	$^nC_{12}H_{25}Cl$	$^nC_{12}H_{25}COOH$	5	89
2	$^nC_{12}H_{25}Br$	$^nC_{12}H_{25}COOH$	2	96
3	$^nC_{12}H_{25}I$	$^nC_{12}H_{25}COOH$	1	97
4	$^nC_{22}H_{45}Br$	$^nC_{22}H_{45}COOH$	6	74
5	$Br(CH_2)_{10}Br$	$HOOC(CH_2)_{10}COOH$	6	76
6	cyclohexyl–Cl	cyclohexyl–COOH	5	45
7	cyclohexyl–Br	cyclohexyl–COOH	6	50
8	cyclohexyl–I	cyclohexyl–COOH	1.5	31
9	Ph–CH$_2$CH$_2$–Cl	Ph–CH$_2$CH$_2$–COOH	4.5	35
10	adamantyl–I	adamantyl–COOH	4.5	-
11	$^nC_{12}H_{25}F$	$^nC_{12}H_{25}COOH$	12 (55 °C)	23

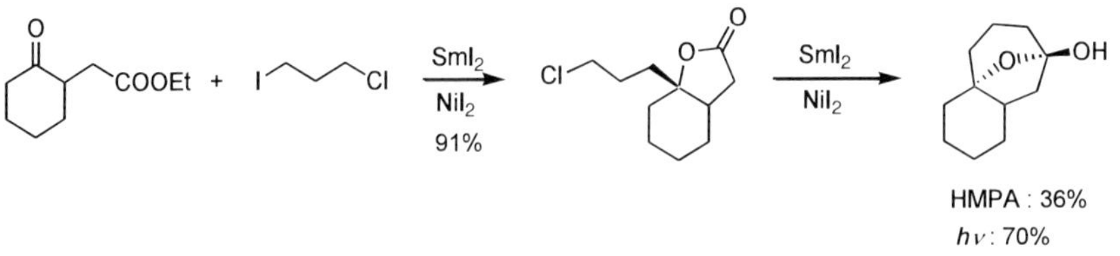

Scheme 46. A tandem reaction of ethyl 2-oxocyclohexaneacetate with 1-chloro-3-iodopropane.

SmI$_2$-$h\nu$ system was successfully applied to an efficient and highly stereo-selective preparation of alkenes with high (Z)-selectivity through β-elimination reaction [79]. When a mixture of O-acetyl chlorohydrins and SmI$_2$ was refluxed in THF upon irradiation with visible light, alkenes were obtained with very high stereoselectivity (Table 23). When this process was carried out in the absence of light, no reaction took place. For example, aliphatic (linear, branched, or cyclic) O-acetyl chlorohydrins afforded the corresponding (Z)-alkenes predominantly (entries 1-6). On the other hand, when aromatic O-acetyl chlorohydrins were used as substrates, (E)-alkenes were obtained with high stereoselectivity (entries 7, 8). The observed stereochemistry of aliphatic alkenes might be explained by a chelation-control model: a six-membered ring intermediate may be generated under the visible light irradiation after metalation of the C-Cl bond (Scheme 48).

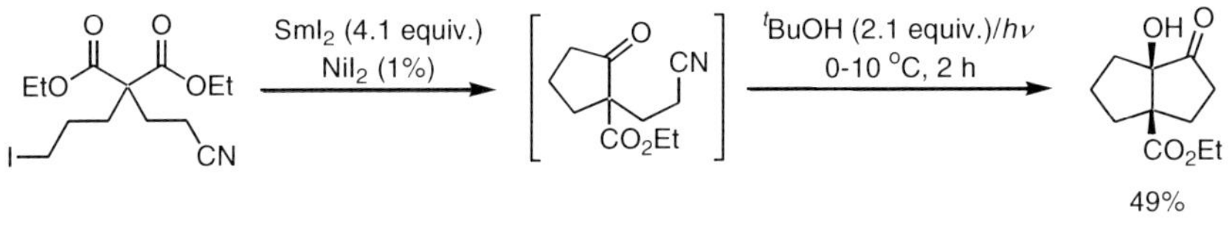

Scheme 47. Cyclization and intramolecular ketone-nitrile coupling.

Hoz et al. demonstrated photoinduced reduction with SmI$_2$ by addition of MeOH [80]. The common denominator for the photostimulated reactions such as Scheme 41 and 43 did not require any bimolecular rate-determining step following the electron transfer. They broadened the scope of the photoinduced reduction with SmI$_2$ to the compounds having double and triple bonds that, under normal conditions, were not amenable to photoirradiation and required bimolecular rate-determining step. Three substrates: naphthalene, diphenylacetylene, and methyl benzoate were studied. With MeOH as the proton donor, naphthalene gave the two isomers in >95% yield (Scheme 49a). These were obtained in a ratio of 7/8= 1.3. The reduction of diphenylacetylene yielded *cis*-and *trans*-stilbene (Scheme 49b). The overall yield was 90% and the *cis*-isomer was obtained, probably as a result of the irradiation of the *trans*-isomer.

Table 23. The highly stereoselective preparation of alkenes

entry	R^1	R^2	Z/E	yield (%)
1	^{n}C$_6$H$_{13}$	Et	>98/2	53
2	^{n}C$_6$H$_{13}$	^{n}Bu	>98/2	72
3	^{n}C$_6$H$_{13}$	^{s}Bu	91/9	51
4	^{n}Pr	^{n}C$_{10}$H$_{21}$	>98/2	65
5	^{n}Pr	Cy	94/6	56
6	^{i}Pr	^{n}C$_8$H$_{17}$	90/10	87
7	Ph	Et	8/92	60
8	Ph	Cy	20/80	67

Scheme 48. Mechanistic proposal for the formation of *cis*-alkenes.

Scheme 49. Photoinduced reductions with SmI$_2$ by addition of MeOH.

The product from methyl benzoate was benzyl alcohol in 80% yield (Scheme 49c). This broadening was facilitated by the fact that methanol complexed efficiently to SmI_2 (Scheme 50). In the next step, excited SmI_2 may either transfer an electron to the substrate or decay to its ground state. Once the radical anion of the substrate is formed, it may be protonated by the complexed MeOH. After that, the protonated substrate gets an electron from another SmI_2-MeOH complex and the bimolecular rate-determining step proceeded.

They also reported the photoirradiated-reduction of nitriles with SmI_2 [81]. In general, the reduction of nitriles with SmI_2 is difficult under regular conditions due to their low electrophilicity [82]. Various nitriles as well as mono- and dicyano compounds could be reduced successfully with SmI_2 under visible light irradiation using MeOH as a proton donor (Scheme 51 and Scheme 52, Table 24).

$$SmI_2 + MeOH \xrightleftharpoons[THF]{} SmI_2(MeOH)_n$$

$$SmI_2(MeOH)_n \xrightleftharpoons{h\nu} SmI_2^*(MeOH)_n$$

$$A + SmI_2^*(MeOH)_n \rightleftharpoons A^{\bullet-} {}^{\bullet}SmI^{3+}I_2(MeOH)_n$$

$$A^{\bullet-} {}^{\bullet}SmI^{3+}I_2(MeOH)_n \longrightarrow HA^{\bullet} + MeO^-Sm^{3+}I_2(MeOH)_{n-1}$$

$$HA^{\bullet} + SmI_2(MeOH)_n \longrightarrow HA^- \longrightarrow Products$$

Scheme 50. Photoinduced reaction of substrate (A) with SmI_2 coordinated with proton donors.

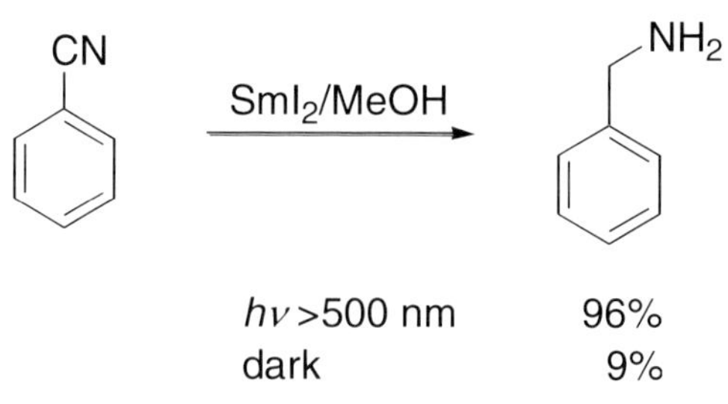

CN

SmI_2/MeOH

NH$_2$

$h\nu$ >500 nm   96%

dark   9%

Scheme 51. Photoirradiated reduction of nitriles with SmI_2.

Scheme 52. Photoirradiated reduction of dicyano compounds with SmI$_2$.

Table 24. Photostimulated reduction of the dicyano compounds

substrate	yield (%)*	
	$h\nu$	dark
o-dicyanobenzene	38	trace
m-dicyanobenzene	11	0
p-dicyanobenzene	68	51

*Quenched after 30 min.

CONCLUSION

A number of activation procedures of SmI$_2$ have been developed so far. As described in this review, the combination of SmI$_2$ and Lewis bases, metals, inorganic salts, H$_2$O, MeOH etc. dramatically enhances the reducing ability of SmI$_2$. Among them, we have developed two important reduction system, i.e., SmI$_2$/Sm and SmI$_2$/$h\nu$.

The SmI$_2$/Sm mixed system is useful for reduction of amides and group 14 heteroatom compounds. The SmI$_2$/$h\nu$ system indicates powerful reducing ability, which is derived from electron-transfer from excited SmI$_2$.

SmI$_2$/$h\nu$ is useful for the reduction of organic chlorides and chalcogenides, and reductive carboxylation of them using CO and CO$_2$, respectively. These novel activation methods mentioned in this review will open up a new utilization of SmI$_2$ in organic synthesis.

REFERENCES

[1]   For reviews concerning samarium diiodide, see for example: (a) Kagan, H. B.; Namy, J. L.; Girard, P. *Tetrahedron* 1981, 37 (Supplement No. 1), 175-180. (b) Kagan, H. B.; Namy, J. L. *Tetrahedron* 1986, 42, 6573-6614. (c) Kagan, H. B.; Sasaki, M.; Collin, J. *Pure Appl. Chem.* 1988, 60, 1725-1730. (d) Molander, G. A. In *Comprehensive Organic Synthesis*; Trost, B. M., Fleming, I.; Eds.; Pergamon: Oxford, 1991; Vol. 1, p 251-282. (e) Molander, G. A. *Chem. Rev.* 1992, 92, 29-68. (f) Molander, G. A. *Org. React.* 1994, 46, 211-367. (g) Murakami, M.; Ito, Y. *J. Organomet. Chem.* 1994, 473, 93-99. (h) Imamoto, T. In *Lanthanides in Organic Synthesis*; Katritzky, A. R.; Meth-Cohn, O.; Rees, C. W.; Eds.; Best Synthetic Methods; Academic Press: San Diego, 1994; pp 21-66. (i) Molander, G. A.; Harris, C. R. *Chem. Rev.* 1996, 96, 307-338. (j) Molander, G. A.; Harris, C. R. *Tetrahedron* 1998, 54, 3321-3354. (k) Kunishima, M.; Tani, S. *J. Synth. Org. Chem. Jpn* 1999, 57, 127-135. (l) Krief, A.; Laval, A. M. *Chem. Rev.* 1999, 99, 745-778. (m) Kagan H. B.; Namy, J. L. In *Lanthanides: Chemistry and Use in Organic Synthesis*; Kobayashi, S.; Ed.; *Topics in Organometallic Chemistry;* Springer: Berlin, Heidelberg, 1999, Vol. 2, pp 155-198. (n) Steel, P. G. *J. Chem. Soc., Perkin Trans. 1* 2001, 2727-2751. (o) Molander G. A. In *Radicals in Organic Synthesis*; Renaud, P.; Sibi, M. P.; Eds.; WILEY-VCH: Weinheim, 2001; Vol. 1, pp 153-182. (p) Sumino, Y.; Ogawa, A. *J. Synth. Org. Chem. Jpn* 2003, 61, 201-210. (q) Kagan, H. B. *Tetrahedron* 2003, 59, 10351-10372. (r) Edmonds, D. J.; Johnston, D.; Procter, D. J. *Chem. Rev.* 2004, 104, 3371-3403. (s) Dahlén, A.; Hilmersson, G. *Eur. J. Inorg. Chem.* 2004, 3393-3403. (t) Gopalaiah, K.; Kagan, H. B. *New J. Chem.* 2008, 32, 607-637; (u) Flowers II, R. A. *Synlett* 2008, 1427-1439. (v) Nicolaou, K. C.; Ellery, S. P.; Chen, J. S. *Angew. Chem. Int. Ed.* 2009, 48, 7140-7165. (w) Procter, D. J.; Flowers, R. A. II.; Skrydstrup, T. *Organic Synthesis using Samarium Diiodide: A Practical Guide;* The Royal Society of Chemistry Publishing: Cambridge, 2010. (x) Szostak, M.; Spain, M.; Procter, D. J. *Chem. Soc. Rev.* 2013, 42, 9155-9183.

[2]   Matignon, C. A.; Caze, E. *Ann. Chim. Phys.* 1906, 8, 417-443.

[3]   Namy, J. -L.; Girard, P.; Kagan, H. B. *Nouv. J. Chim.* 1977, 1, 5-7.(b) Girard, P.; Namy, J.-L.; Kagan, H. B. *J. Am. Chem. Soc.* 1980, 102, 2693-2698.

[4]   Imamoto, T.; Ono, M.; *Chem. Lett.* 1987, 501-502.

[5]   Akane, N.; Kanagawa, Y.; Nishiyama, Y.; Ishii, Y.; *Chem. Lett.* 1992, 2431-2434.

[6]   Concellón, J. M.; Rodríguez-Solla, H.; Bardales, E.; Huerta, M. *Eur. J. Org. Chem.* 2003, 1775-1778. (b) Teprovich, J. A. Jr.; Antharjanam, P. K. S.; Prasad, E.; Pesciotta, E. N.; Flowers, R. A. II. *Eur. J. Inorg. Chem.* 2008, 5015-5019.

[7]   Dahlén, A.; Hilmersson, G. *Eur. J. Inorg. Chem.* 2004, 3020-3024.

[8]   Szostak, M.; Spain, M.; Procter, D. J. *J. Org. Chem.* 2012, 77, 3049-3059.

[9]   Voltrova, S.; Srogl, J. *Synlett* 2013, 24, 394-396.

[10]   Johnson, D. A. *J. Chem. Soc., Dalton Trans.* 1974, 1671-1675.

[11]   Shabangi, M.; Flowers II, R. A. *Tetrahedron Lett.* 1997, 38, 1137-1140. (b) Enemaerke, R. J.; Daasbjerg, K.; Skrydstrup, T. *Chem. Commun.* 1999, 343-344.

[12]   Enholm E. J.; Jiag, I.; Abboud, K. *J. Org. Chem.* 1993, 58, 4061-4069. (b) Mukaiyama, T.; Arai, H.; Shiina, I. *Chem. Lett.* 2000, 580-581. (c) Jacobsen M. F.; Turks, M.; Hazell, R.; Skrydstrup, T. *J. Org. Chem.* 2002, 67, 2411-2417. (d) McKerlie, F.; Rudkin, I. M.; Wynne G.; Procter, D. J. *Org. Biomol. Chem.* 2005, 3, 2805-2816. (e) Ogawa, Y.; Kuroda, K.; Mukaiyama, T. *Chem. Lett.* 2005, 34, 698-699.

[13]   Molander, G. A.; Harring, L. S. *J. Org. Chem.* 1990, 55, 6171-6176. (b) Molander, G. A.; McKie, J. A. *J. Org. Chem.* 1991, 56, 4112-4120. (c) Curran, D. P.; Totleben, M. J. *J. Am. Chem. Soc.* 1992, 114, 6050-6058. (d) Curran, D. P.; Fevig, T. L.; Jasperse, C. P.; Totleben, M. J. *Synlett* 1992, 943-961. (e) Ito, Y.; Takahashi, K.; Nagase, H.; Honda, T. *Org. Lett.* 2011, 13, 4640-4643.

[14]   Molander, G. A.; Etter, J. B. *J. Am. Chem. Soc.* 1987, 109, 6556-6558. (b) Moriya, T.; Handa, Y.; Inanaga, J.; Yamaguchi, M. *Tetrahedron Lett.* 1988, 29, 6947-6948. (c) Fukuzawa, S.; Matsuzawa, H.; Yoshimitsu, S. *J. Org. Chem.* 2000, 65, 1702-1706. (d) Concellón, J. M.; Concellón, C.; Mejica, C. *J. Org. Chem.* 2005, 70, 6111-6113. (e) Utimoto, K.; Matsui, T.; Takai, T. Matsubara, S. *Chem. Lett.* 1995, 197-198. (f) Otaka, A.; Watanabe, J.; Yukimasa, A.; Sakai, Y.; Watanabe, H.; Kinoshita, T.; Oishi, S.; Tamamura, H.; Fujii, N. *J. Org. Chem.* 2004, 69, 1634-1645. (g) Nelson, C. G.; Burke, Jr, T. R. *J. Org. Chem.* 2012, 77, 733-738.

[15]   Namy, J. L.; Souppe, J.; Kagan, H. B. *Tetrahedron Lett.* 1983, 24, 765-766. (b) Nishiyama, Y.; Shinomiya, E.; Kimura, S.; Itoh, K.; Sonoda, N. *Tetrahedron Lett.* 1998, 39, 3705-3708. (c) Mashima, K.; Oshiki, T.; Tani, K. *J. Org. Chem.* 1998, 63, 7114-7116. (d) Fukuzawa, S.;

Tsuchimoto, T.; Kanai, T. *Chem. Lett.* 1994, 1981-1984. (e) Ogawa, A.; Takeuchi, H.; Hirao, T. *Tetrahedron Lett.* 1999, 40, 7113-7114. (f) Masson, G.; Py, S.; Vallée, Y. *Angew. Chem. Int. Ed.* 2002, 41, 1772-1775. (g) Zhong, Y. -W.; Dong, Y. -Z.; Fang, K.; Izumi, K.; Xu, M. -H.; Lin, G. -O. *J. Am. Chem. Soc.* 2005, 127, 11956-11957. (h) Ebran, J. -P.; Hazell, R. G.; Skrydstrup, T. *Chem. Commun.* 2005, 5402-5404.

[16] Molander, G. A.; McKie, J. A. *J. Org. Chem.* 1992, 57, 3132-3139. (b) Kakiuchi, K.; Fujioka,Y.; Yamamura, H.; Tsutsumi, K.; Morimoto, T.; Kurosawa, H. *Tetrahedron* 2001, 42, 7595-7598. (c) Johnston, D.; Couché, E.; Edmonds, D. J.; Muir, K. W.; Procter, D. J. *Org. Biomol. Chem.* 2003, 1, 328-337. (d) Hutton, T. K.; Muir, K. W.; Procter, D. J. *Org. Lett.* 2003, 5, 4811-4814. (e) Molander, G. A.; Czakó, B.; Rheam, M. *J. Org. Chem.* 2007, 72, 1755-1764. (f) Molander, G. A.; George, K. M.; Monovich, L. G. *J. Org. Chem.* 2003, 68, 9533-9540.(g) Xu, T.; Li, C. C.; Yang, Z. *Org. Lett.* 2011, 13, 2630-2633.

[17] Tabuchi, T.; Kawamura, K.; Inanaga, J., Yamaguchi, M. *Tetrahedron Lett.* 1986, 27, 3889-3890. (b) Kito, M.; Sakai, T.; Shirahama, H.; Miyashita, M.; Matsuda, F. *Synlett* 1997, 219-220. (c) Matsuda, F.; Kito, M.; Sakai, T.; Okada, N.; Miyashita, M.; Shirahama, H. *Tetrahedron* 1999, 55, 14369-14380. (d) Monovich, L. G.; Huèrou, Y. L.; Rönn, M.; Molander, G. A. *J. Am. Chem. Soc.* 2000, 122, 52-57.

[18] Mukaiyama, T.; Shiina, I.; Iwadate, H.; Saitoh, M.; Nishimura, T.; Ohkawa, N.; Sakoh, H.; Nishimura, K.; Tani, Y.; Hasegawa, M.; Yamada, K.; Saitoh, K. *Chem. Eur. J.* 1999, 5, 121-161. (b) Edmonds, D. J.; Johnston, D.; Procter, D. J. *Chem. Rev.* 2004, 104, 3371-3404. (c) Park, J.; Kim, B.; Kim, H.; Kim, S.; Kim, D. *Angew. Chem. Int. Ed.* 2007, 46, 4726-4728. (d) Fukui, H.; Shiina, I. *Org. Lett.* 2008, 10, 3153-3156. (e) Sono, M.; Doi, N.; Yoshino, E.; Onishi, S.; Fujii, D.; Tori, M. *Tetrahedron Lett.* 2013, 54, 1947-1950.

[19] (a) Du, X.; Armstrong, R. W. *J. Org. Chem.* 1997, 62, 5678-5679. (b) Myers, R. M.; Langston, S. P.; Conway, S. P.; Abell, C. *Org. Lett.* 2000, 2, 1349-1352. (c) Kerrigan, N. J.; Hutchison, P. C.; Heightman, T. D.; Procter, D. J. *Chem. Commun.* 2003, 1402-1403. (d) Kerrigan, N. J.; Hutchison, P. C.; Heightman, T. D.; Procter, D. J. *Org. Biomol. Chem.* 2004, 2476-2482.

[20] Takeuchi, S.; Nakamura, Y.; Ohgo, Y.; Curran, D. P. *Tetrahedron Lett.* 1998, 39, 8691-8694. (b) Vogel, J. C.; Butler, R.; Procter, D. J. *Tetrahedron* 2008, 64, 11876-11883.

[21] Ogawa, A.; Takami, N.; Sekiguchi, M.; Ryu, I.; Kambe, N.; Sonoda, N. *J. Am. Chem. Soc.* 1992, 114, 8729-8730. (b) Ogawa, A.; Nanke, T.; Takami, N.; Sumino, Y.; Ryu, I.;Sonoda, N. *Chem. Lett.* 1994, 379-380. (c) Ogawa, A.; Nanke, T.; Takami, N.;Sekiguchi, M.;Kambe, N.;Sonoda, N. *Appl. Organomet. Chem.* 1995, 9, 461-466. (d) Ogawa, A.;Takami, N.; Nanke, T.; Ohya, S.; Hirao, T.; Sonoda, N. *Tetrahedron* 1997, 53, 12895-12902. (e) Kamiya, I. Iida, K.; Harato, N.; Li, Z.; Tomisaka, Y.; Ogawa, A. *J. Alloys Compd.* 2006, 408-412, 437-440. (f) Li, Z.; Iida, K.; Tomisaka, Y.; Yoshimura, A.; Hirao, T.; Nomoto, A.; Ogawa, A. *Organometallics* 2007, 26, 1212-1216.

[22] Ogawa, A.; Sumino, Y.; Nanke, T.; Ohya, S.; Sonoda, N.; Hirao, T. *J. Am. Chem. Soc.* 1997, 119, 2745-2746. (b) Ogawa, A.; Sumino, Y.; Nanke, T.; Ryu, I.; Kambe, N.; Sonoda, N. *Rare Earths* 1995, 338-339. (c) Ogawa, A.; Hirao, T.; Sumino, Y.; Sonoda, N. *Rare Earths* 1996, 298-299.(d) Imamoto, T.; Tawarayama, Y.; Kusumoto, T.; Yokoyama, M. *J. Synth. Org. Chem. Jpn.* 1984, 42, 143-152. (e) Skene, W. G.; Scaiano, J. C.; Cozens, F. L. *J. Org. Chem.* 1996, 61, 7918-7921. (f) Ogawa, A.; Ohya, S.; Hirao, T. *Chem. Lett.* 1997, 275-276. (g) Ogawa, A.; Ohya, S.; Sumino, Y.; Sonoda, N.; Hirao, T. *Tetrahredron Lett.* 1997, 38, 9017-9018. (h) Ogawa, A.; Ohya, S.; Doi, M.; Sumino, Y.; Sonoda, N.; Hirao, T. *Tetrahedron Lett.* 1998, 39, 6341-6342. (i) Sumino, Y.; Harato, N.; Tomisaka, Y.; Ogawa, A. *Tetrahedron* 2003, 59, 10499-10508. (j) Tomisaka, Y.; Harato, N.; Sato, M.; Nomoto, A.; Ogawa, A. *Bull. Chem. Soc. Jpn.* 2006, 79, 1444-1446. (k) Nomoto, A.; Kojo, Y.; Shiino, G.; Tomisaka, Y.; Mitani, I.; Tatsumi, M.; Ogawa, A. *Tetrahedron Lett.* 2010, 51, 6580-6583.

[23] Inanaga, J.; Ishikawa, M.; Yamaguchi, M. *Chem. Lett.* 1987, 1485-1486.

[24] Molander G. A.; McKie, J. A *J. Org. Chem.* 1992, 57, 3132-3139.

[25] Hou, Z.; Wakatsuki, Y. *J. Chem. Soc., Chem. Commun.* 1994, 1205-1206.

[26] Hasegawa, E.; Curran, D. P. *Tetrahedron Lett.* 1993, 34, 1717-1720.

[27] Jackson, H.; Jones, A.; Cooper, E. *J. Reprod. Fertil.* 1969, 20, 263-269.

[28] Fevig, T. L.; Elliott, R. L.; Curran, D. P. *J. Am. Chem. Soc.* 1988, 110, 5064-5067. (b) Hasegawa, E.; Curran, D. P. *J. Org. Chem.* 1993, 58, 5008-5010.

[29] Shabangi, M.; Sealy, J. M.; Fuchs, J. R.; Flowers II, R. A. *Tetrahedron Lett.* 1998, 39, 4429-4432. (b) Kuhlman, M. L.; Flowers II, R. A. *Tetrahedron, Lett.* 2000, 41, 8049-8052.

[30] Hojo, M.; Aihara, H.; Hosomi, A. *J. Am. Chem. Soc.* 1996, 118, 3533-3534.

[31] Kumaran, R.; Bruedgam, I.; Reissig, H. U. *Synlett* 2008, 991-994.

[32] Cabri, W.; Candiani, I.; Colombo, M.; Franzoi, L.; Bedeschi, A. *Tetrahedron, Lett.* 1995, 36, 949-952.

[33] Prasad, E.; Knettle, B. W.; Flowers II, R. A. *J. Am. Chem. Soc.* 2002, 124, 14663-14667.

[34] McDonald, C. E.; Ramsey, J. R.; Sampsell, D. G.; Anderson, L. A.; Krebs, J. E.; Smith, S. N. *Tetrahedron* 2013, 69, 2947-2953.

[35] McDonald, C. E.; Ramsey, J. D.; Sampsell, D. G.; Butler, J. A.; Cecchini, M. R. *Org. Lett.* 2010, 12, 5178-5181. (b) Berndt, M.; Hölemann, A.; Niermann, A.; Bentz, C.; Zimmer, R.; Reissig, H. –U. *Eur. J. Org. Chem.* 2012, 1299-1302.

[36] Miller, R. S.; Sealy, J. M.; Shabangi, M.; Kuhlman, M. L.; Fuchs, J. R.; Flowers II, R. A. *J. Am. Chem. Soc.* 2000, 122, 7718-7722. (b) Prasad, E.; Flowers II, R. A. *J. Am. Chem. Soc.* 2002, 124, 6859-6899.

[37] Hélion, F.; Namy, J. L. *J. Org. Chem.* 1999, 64, 2944-2946. (b) Lannou, M. I.; Hélion, F.; Namy, J. L. *Tetrahedron* 2003, 59, 10551-10565.

[38] Tomisaka, Y.; Yoshimura, A.; Nomoto, A.; Sonoda, N.; Ogawa, A. *Res. Chem. Intermed.* 2013, 39, 43-48.

[39] Nomura, R.; Matsuno, T.; Endo, T. *J. Am. Chem. Soc.* 1996, 118, 11666-11667.

[40] Jong, S. J.; Fang, J. M. *J. Org. Chem.* 2001, 66, 3533-3537. (b) Xu, X. L.; Zhang, Y. M. *Chin. J. Chem.* 2002, 20, 1463-1465. (c) Thiessen, W.; Wolff, T. *Designed Monomers and Polymers*, 2005, 8, 397-407. (d) Aspinall, H. C.; Greeves, N.; Valla, C. *Org. Lett.* 2005, 7, 1919-1922. (e) Orsini, F.; Lucci, E. M. *Tetrahedron Lett.* 2005, 46, 1909-1911.

[41] Corey, E. J.; Zheng, Z. *Tetrahedron Lett.* 1997, 38, 2045-2048.

[42] Girard, P.; Namy, J. L.; Kagan, H. B. *J. Am. Soc. Chem.* 1980, 102, 2693-2698. (b) Molander, G. A.; Etter, J. B. *J. Org. Chem.* 1986, 51, 1778-1786. (c) Kagan, H. B. *J. Alloys Compd.* 2006, 408-412, 421-426.

[43] Machrouhi, F.; Hamann, B.; Namy, J. L.; Kagan, H. B. *Synlett* 1996, 633-634.

[44] Machrouhi, F.; Namy, J. L.; Kagan, H. B. *Tetrahedron Lett.* 1997, 38, 7183-7186.

[45] Machrouhi, F.; Namy, J. L. *Tetrahedron Lett.* 1999, 40, 1315-1318.

[46] Molander, G. A.; Harris, C. R. *J. Org. Chem.* 1997, 62, 7418-7429. (b) Molander, G. A.; Jean, Jr. D. J. *J. Org. Chem.* 2002, 67, 3861-3865.

[47] Kang, H. Y.; Song, S. E. *Tetrahedron Lett.* 2000, 41, 937-939.

[48]   Molander, G. A.; Huérou, Y. L.; Brown, G. A. *J. Org. Chem.* 2001, 66, 4511-4516.

[49]   Tabuchi, T.; Inanaga, J.; Yamaguchi, M. *Tetrahedron Lett.* 1986, 27, 601-602. (b) Tabuchi, T.; Inanaga, J.; Yamaguchi, M. *Tetrahedron Lett.* 1986, 27, 1195-1196. (c) Tabuchi, T.; Inanaga, J.; Yamaguchi, M. *Tetrahedron Lett.* 1986, 27, 5237-5240. (d) Yoshida, A.; Mikami, K.; *Synlett* 1997, 1375-1376.

[50]   Totleben, M. J.; Curran, D. P.; Wipf, P. *J. Org. Chem.* 1992, 57, 1740-1744. (b) Berkowitz, W.; F.; Wu, Y. *Tetrahedron Lett.* 1997, 38, 3171-3174.

[51]   Inanaga, J.; Yokoyama, Y.; Baba, Y.; Yamaguchi, M. *Tetrahedron Lett.* 1991, 32, 5559-5562.

[52]   Matsubara, S.; Horiuchi, M.; Takai, K.; Utimoto, K. *Chem. Lett.* 1995, 259-260. (b) Matsubara, S.; Yoshioka, M.; Utimoto, *Angew. Chem. Int. Ed.* 1997, 36, 617-618. (c) Utimoto, K.; Matsubara, S. *J. Synth. Org. Chem. Jpn.* 1998, 56, 908-918.

[53]   Kamochi, Y.; Kudo, T. *Chem. Lett.* 1991, 893-896. (b) Kamochi, Y.; Kudo, T. *Tetrahedron Lett.* 1991, 32, 3511-3514. (c) Kamochi, Y.; Kudo, T. *Bull. Chem. Soc. Jpn* 1992, 65, 3049-3054.

[54]   Kamochi, Y.; Kudo, T. *Tetrahedron Lett.* 1994, 35, 4169-4172.

[55]   Kamochi, Y.; Kudo, T. *Tetrahedron* 1992, 48, 4301-4312.

[56]   Studer, A.; Curran, D. P. *Synlett* 1996, 3, 255-257.

[57]   Ohmori, K.; Kitamura, M.; Ishikawa, Y.; Kato, H.; Oorui, M.; Suzuki, K. *Tetrahedron Lett.* 2002, 43, 7023-7026.

[58]   Liu, X.-K.; Qiu, S.; Xiang, Y.-G.; Ruan, Y.-P.; Zheng, X.; Huang, P.-Q. *J. Org. Chem.* 2011, 76, 4952-4963.

[59]   Inanaga, J.; Katsuki, J.; Yamaguchi, M. *Chem. Lett.* 1991, 1025-1026.

[60]   Hasegawa, E.; Curran, D. P. *J. Org. Chem.* 1993, 58, 5008-5010.

[61]   Kamochi, Y.; Kudo, T. *Chem. Lett.* 1993, 1495-1498. (b) Kamochi, Y.; Kudo, T. *Heterocycles* 1993, 36, 2383-2396.

[62]   Dahlén, A.; Hilmersson, G. *Chem. Eur. J.* 2003, 9, 1123-1128. (b) Dahlén, A.; Hilmersson, G.; Knettle, B. W.; Flowers, II, R. A. *J. Org. Chem.* 2003, 68, 4870-4875. (c) Dahlén, A.; Sundgren, A.; Lahmann, M.; Oscarson, S.; Hilmersson, G. *Org. Lett.* 2003, 5, 4085-4088. (d) Davis, T. A.; Chopade, P. R.; Hilmersson, G.; Flowers, II, R. A. *Org. Lett.* 2005, 7, 119-122. (e) Dahlén, A.; Hilmersson, G. *Tetrahedron Lett.* 2003, 44, 2661-2664. (f) Dahlén, A.; Nilsson, Å.; Hilmersson, G. *J. Org. Chem.* 2006, 71, 1576-1580. (g) Ankner, T.; Hilmersson, G. *Org. Lett.*

2009, 11, 503-506. (h) Wettergren, J.; Ankner, T.; Hilmersson, G. *Chem. Commun.* 2010, 46, 7596-7597.

[63]   Szostak, M.; Spain, M.; Procter, D. J. *Chem. Commun.* 2011, 47, 10254-10256. (b) Szostak, M.; Spain, M.; Procter, D. J. *Org. Lett.* 2012, 14, 840-843. (c) Szostak, M.; Sautier, B.; Spain, M.; Procter, D. J. *Org. Lett.* 2014, 16, 1092-1095.

[64]   Szostak, M.; Sautier, B.; Spain, M.; Behlendorf, M.; Procter, D. J. *Angew. Chem. Int. Ed.* 2013, 52, 12559-12563.

[65]   Molander, G. A.; Hahn, G. *J. Org. Chem.* 1986, 51, 1135-1138.

[66]   Keck, G. E.; Wager, C. A.; Sell, T.; Wager, T. T. *J. Org. Chem.* 1999, 64, 2172-2173. (b) Keck, G. E.; Truong, A. P. *Org. Lett.* 2002, 4, 3131-3134. (c) Keck, G. E.; Giles, R. L.; Cee, V. J.; Wager, C. A. Yu, T.; Kraft, M. B. *J. Org. Chem.* 2008, 73, 9675-9691.

[67]   Dahlén, A.; Hilmersson, G. *Tetrahedron Lett.* 2001, 42, 5565-5569.

[68]   Teprovich, Jr. J. A.; Balili, M. N.; Pintauer, T.; Flowers, II, R. A. *Angew. Chem. Int. Ed.* 2007, 46, 8160-8163.

[69]   Hutton, T. K.; Muir, K. W.; Procter, D. J. *Org. Lett.* 2003, 5, 4811-4814.

[70]   Kleiner, G.; Tarnopolsky, A.; Hoz, S. *Org Lett.* 2005, 7, 4197-4200. (b) Farran, H.; Hoz, S. *J. Org. Chem.* 2009, 74, 2075-2079. (c) Upadhyay, S. K.; Hoz, S. *J. Org. Chem.* 2011, 76, 1355-1360.

[71]   Okaue, Y.; Isobe, T. *Inorg. Chim. Acta,* 1988, 144, 143-146.

[72]   Prasad, E.; Knettle, B. W.; Flowers II, R. A. *Chem. Eur. J.* 2005, 11, 3105-3112.

[73]   Lusztyk, J.; Maillard, B.; Deycard, S.; Lindsay, D. A.; Ingold, K. U. *J. Org. Chem.* 1987, 52, 3509-3514.

[74]   Hasegawa, E.; Curran, D. P. *Tetrahedron, Lett.* 1993, 34, 1717-1720.

[75]   Russell, G. A.; Tashtoush, H. *J. Am. Chem. Soc.* 1983, 105, 1398-1399.

[76]   Souppe, J.; Namy, J. L.; Kagan, H. B. *Tetrahedron Lett.* 1984, 25, 2869-2872. (b) Collin, J.; Dallemer, F.; Namy, J. L.; Kagan, H. B. *Tetrahedron Lett.* 1989, 30, 7407-7410.

[77]   Molander, G. A.; Alonso-Alija, C. *J. Org. Chem.* 1998, 63, 4366-4373. (b) Molander, G. A.; Sono, M. *Tetrahedron* 1998, 54, 9289-9302.

[78]   Molander, G. A.; Wolfe, C. N. *J. Org. Chem.* 1998, 63, 9031-9036.

[79]   Concellón, J. M.; Rodríguez-Solla, H.; Simal, C.; Huerta, M. *Org. Lett.* 2005, 7, 5833-5835. (b) Concellón, J. M.; Rodríguez-Solla, H.; Simal, C.; Santos, D.; Paz, N. R. *Org. Lett.* 2008, 10, 4549-4552.

[80]   Amiel-Levy, M.; Hoz, S. *Chem. Eur. J.* 2010, 16, 805-809.

[81]   Rao, C. N.; Hoz, S. *J. Org. Chem.* 2012, 77, 4029-4034.

[82]   Souppe, J.; Danon, L.; Namy, J. L.; Kagan, H. B. *J. Organomet. Chem.* 1983, 250, 221-236.

In: Samarium   ISBN: 978-1-63321-045-5
Editor: Kaitlyn R. Danford   © 2014 Nova Science Publishers, Inc.

Chapter 4

COMPARISON OF Sm(III) AND Cr(VI) IONS FOR VISIBLE LIGHT INDUCED REDUCTION IN METHANOL BY HYBRID SYSTEMS OF CHIRAL SCHIFF BASE Cu(II) COMPLEXES AND TiO_2

*Naoki Yoshida[1] and Takashiro Akitsu[1]**

[1]Department of Chemistry, Faculty of Science,
Tokyo University of Science, Kagurazaka,
Shinjuku-ku Tokyo, Japan

ABSTRACT

This paper reports on effective visible light induced reduction of Cr(VI) ion chiral Schiff base Cu(II) complex and TiO_2 in non-aqueous organic solvent (methanol) as a model system for environmental clean-up. By using several new and known chiral Schiff base Cu(II) complexes, hybrid systems of the Cu(II) complexes and anatase TiO_2 were prepared and reduction of Cr(VI) ion after visible light irradiation.

The Cu(II) complexes were synthesized in a common procedure and characterized by means of elemental analysis, IR, UV-vis, CD, and ESR spectra. Hybrid systems of the Cu(II) complex and TiO_2 were prepared as methanol suspension. After visible light irradiation, Cr(VI) ion was

---

* E-mail: akitsu@rs.kagu.tus.ac.jp.

reduced successfully, which was confirmed by diphenylthiocarbazide analysis method. To our knowledge, this visible light reaction in non-aqueous organic solvents can be observed for the first time. However, reduction of Cr(VI) ion could not be observed for only a part of components selected of the hybrid systems. Although similar reaction was tested for Sm(III) ion instead of Cr(VI) ion for comparison, no change could be detected in emission as well as absorption spectra in the case of Sm(III) ion due to inappropriate redox potential.

INTRODUCTION

Recently, we have reported photo-induced reduction of copper(II) complexes with TiO_2 by UV light irradiation [1-9]. Especially, some Schiff base copper(II) complexes having L-amino acid derivatives have been investigated by means of spectroelectrochemical or electrochemical measurements for photo-induced electron transfer reactions of some complexes by TiO_2 to give rise to copper(I) species after UV light irradiation Beside them, in this course work, we have found that reaction conditions may depend on chemical structure of copper(II) complexes [1], complicate reaction mechanism was proposed based on experimental results [2], redox potential may be controlled by ligands [3-6], and azo-dyes can act as fluorescence prove for reduced copper(I) species [7-9].

By the way, UV light-induced reduction of metal ions by TiO_2 has also been reported [10, 11]. The environmental standard and the effluent standard are established by government for Cr(VI) ion in Japan because it is poisonous due to a strong oxidizing agent. Though most of studies on reduction of Cr(VI) ion are carried out in aqueous solutions, few studies in organic solvents have been reported to date. It should be noted that Sm(III) ion is also tested for comparison because reduced Sm(II) ion is known to be relatively stable among lanthanide ions.

Herein, we have attempted *visible light*-induced reduction of Cr(VI) to Cr(III) in *organic solvent* (methanol) effectively by using hybrid system composed of TiO_2 and chiral Schiff base copper(II) complexes, known ones abbreviated as *CuEBSa, CuEBDC*, and *CuEBtB* (Scheme 1) [1] and new ones abbreviated as *CuLSa*, and *CuLDC* (Scheme 2). Interestingly, as for two components systems of TiO_2 and copper(II) complexes, the former cannot be reduced by TiO_2, while the latter can be reduced by TiO_2. Concentration of residual Cr(VI) ions is measured with a diphenylcarbazide method, which can indicate that bottom sediment of Tama river (Tokyo, Japan) did contain Cr(VI) ions (less than 0.005 mg/L confirmed with an ICP method).

	R
CuEBSa	H
CuEBDC	Cl
CuEBtB	tert-buthyl

Scheme 1. Molecular structures of *CuEBSa*, *CuEBDC*, and *CuEBtB* [1].

	R
CuLSa	H
CuLDC	Cl

Scheme 2. Molecular structures of *CuLSa* and *CuLDC*.

EXPERIMENTAL SECTION

Materials

All reagents and solvents were commercially available and were used as purchased without further purification. Amine ligands were prepared by treatment of L-glutamic acid (10 mmol) and benzylalchol (12 mmol) in 60 % H_2SO_4 at 343 K for 4h, and after vacuum concentration the product was neutralized to give rise to L-glutamic acid ester. *CuEBSa, CuEBDC,* and *CuEBtB* were prepared according to the literature procedure [1]. 1 mM Cr(VI) aqueous solution was prepared from $K_2Cr_2O_7$.

Preparations of CuLSa and CuLDC

To a methanol solution of L-leucine (2 mmol) and the corresponding salicylaldehyde (*CuLSa*) or 3,5-dichloro salicylaldehyde (CuLDC) (2 mmol) at 318 K, $Cu(CH_3COO)_2 \cdot H_2O$ (1 mmol) were added and stirred for 2 h. Then imidazol (1 mmol) was added and stirred for 1 h.

CuLSa. Yield 54.8 %. Anal. Found: C, 53.05; H, 5.08; N, 11.52 %. Calc. for $C_{16}H_{18}CuN_3O_3$: C, 52.81; H, 4.99; N, 11.55 %. IR (KBr) 1635 cm^{-1} (C=N). UV-vis (diffuse reflectance) 389 nm ($\pi-\pi^*$), 610 nm (d-d). CD (KBr) 347 nm, 415 nm. (absorption, methanol) 370 nm ($\pi-\pi^*$), 640 nm (d-d). CD (methanol) 370 nm ($\pi-\pi^*$). ESR(methanol, 77K) g = 2.044.

CuLDC. Yield 52.8 %. Anal. Found: C, 44.60; H, 3.70; N, 9.69 %. Calc. for $C_{16}H_{16}CuCl_2N_3O_3$: C, 44.41; H, 3.73; N, 9.71 %. IR (KBr) 1636 cm^{-1} (C=N). UV-vis (diffuse reflectance) 420 nm ($\pi-\pi^*$), 606 nm (d-d). CD (KBr) 289 nm, 416 nm. (absorption, methanol) 385 nm ($\pi-\pi^*$), 640 nm (d-d). CD (methanol) 390 nm ($\pi-\pi^*$). ESR(methanol, 77K) g = 2.041.

Physical Measurements

Elemental analyses (C, H, N) were carried out with a Perkin-Elmer 2400II CHNS/O analyzer at Tokyo University of Science. Infrared spectra were recorded a JASCO FT-IR 4200 spectrophotometer equipped with a polarizer in the range of 4000-400 cm^{-1} at 298 K.

Absorption electronic spectra were measured on a JASCO V-570 spectrophotometer in the range of 900-200 nm at 298 K. Circular dichroism (CD) spectra were measured on a JASCO J-725 spectropolarimeter in the range of 800-200 nm at 298 K. X-band ESR spectra of solutions were measured with a JEOL JES-FA200 spectrometer at 77K. Spectroelectrochemical measurements were carried out on a BAS SEC2000-UV/CVIS and ALS2323 system with Ag/AgCl electrodes in aqueous solutions. UV and visible light source used was Hayashi LA-310UV and LA-251Xe, respectively with visible ($\lambda > 350$ nm) or UV ($\lambda < 350$ nm) cut filters.

RESULTS AND DISCUSSION

Characterization of copper(II) complexes. As mentioned in the experimental sections, the CD and UV-vis spectra in methanol solutions for *CuEBSa, CuEBDC, CuEBtB, CuLSa,* and *CuLSa,* have been recorded (d-d band at about 650 nm and $\pi-\pi^*$ band at around 250-400 nm), X-band EPR spectra (g = 2.03 − 2.05 and their hyperfine splitting) which are normal values for the analogous (mononuclear) copper(II) complexes except for slight shift due to different ligands (in the region of about 500-800 nm).

In contrast to our related copper(II) complexes of four-coordinated ones, the results of elemental analysis for *CuEBSa, CuEBDC,* and *CuEBtB* in the solid states suggested to be three-coordinated ones or four-coordinated one with additional oxygen coordination by ligands of neighboring molecules. Because ESR spectra for *CuEBSa, CuEBDC,* and *CuEBtB* in the solutions exhibited similar features to analogous four-coordinated copper(II) complexes having solvent molecules as a ligand, *CuEBSa, CuEBDC,* and *CuEBtB* could be existed to be four-coordinated ones in the solutions.

Although appropriate single crystals of *CuLSa* and *CuLSa* suitable for X-ray crystallography could not be obtained unfortunately, their structures could be estimated as Scheme 2 because of their composition. Crystal structure of analogous salicylaldehyde copper(II) complexes having L-alanine instead of L-leucine moiety was determined to afford a four-coordinated [CuN$_2$O$_2$] environment (tridentate NO$_2$ ligand and one of the two coordination N atom is from imidazol ligand).

Therefore, identical four-coordinated [CuN$_2$O$_2$] environment was also estimated for *CuLSa* and *CuLSa* due to similar composition to the known complex.

Hybrid system of Cu(II) complex and TiO$_2$. Our previous studies elucidated that possibility of UV light-induced reduction of Cu(II) complex to Cu(I) complex by TiO$_2$ depends on chemical structures of copper(II) complexes.

Some copper(II) complexes such as *CuEBSa, CuEBDC,* and *CuEBtB* could not exhibit the photo-reduction reaction. On the other hand, *CuLSa* and *CuLSa* could exhibit the photo-reduction like many other analogous copper(II) complexes. It should be noted that copper(II) complexes incorporating imidazol ligands result in other reaction mechanism against copper(II) complexes having solvent ligands. Reduction to Cu(I) ion to form three-coordinated complex accompanying with elimination of an imidazol ligand.

Although oxidation to Cu(II) ion is reversible, coordination of an imidazol ligand is irreversible even after recovering a four-coordinated $[Cu^{II}N_2O_2]$ environment (tridentate NO_2 ligand and one of the two coordination N atom is from *solvent* ligand).

Visible light could not excite electrons and could not employ for the photo-reduction reaction mentioned above and it must be low-efficient reaction. However, besides metal ions to be reduced, mixing visible-light absorbing metal complex into the hybrid system may be useful strategy for employing visible light for the photo-reduction reaction.

Photoreduction of Cr(VI) by hybrid systems. Figures 1-5 exhibit spectral changes of of hybrid sytems of copper(II) complexes (*CuEBSa*, *CuEBDC*, *CuEBtB*, *CuLSa*, and *CuLDC*, respectively), TiO_2, and Cr(VI) after visible light irradiation for 0, 20, 40, and 60 nm.

The hybrid systems were prepared by adding 1 mL of 1 mM Cr(VI) $(K_2Cr_2O_7)$ aqueous solution to 25 mL of methanol solution (by mixing 1 mM methanol solution of copper(II) complexes and 0.3 mM methanol suspension of TiO_2).

Soon after preparation of the hybrid systems of three components, visible light was irradiated. For all the hybrid systems, characteristic features of UV-vis spectra associated with copper(II) complexes were maintained, which indicated copper(II) complexes did not decompose even after light irradiation and were stable against except for photo-induced reaction (in other words merely mixing solutions).

After reaction for 0, 20, 40, and min, concentration of Cr(VI) ion was measured with a diphenylcarbazide method using UV-vis spectra. Although all five hybrid systems could be observed decreasing concentration of Cr(VI) ion after the photo-induced reactions, the system containing *CuLDC* exhibited the smallest decreasing reaction of Cr(VI) ions among them. In order to qualitative comparison of the degree of Cr(VI) concentration (namely residual Cr(VI) concentration), equation (1) is employed:

$$\Delta\alpha \text{ rate} = (\text{Abs}(540 \text{ nm}) - \text{BG}) / (\text{Abs}(540 \text{ nm at 0 min}) - \text{BG})) \qquad (1)$$

where BG means background values.

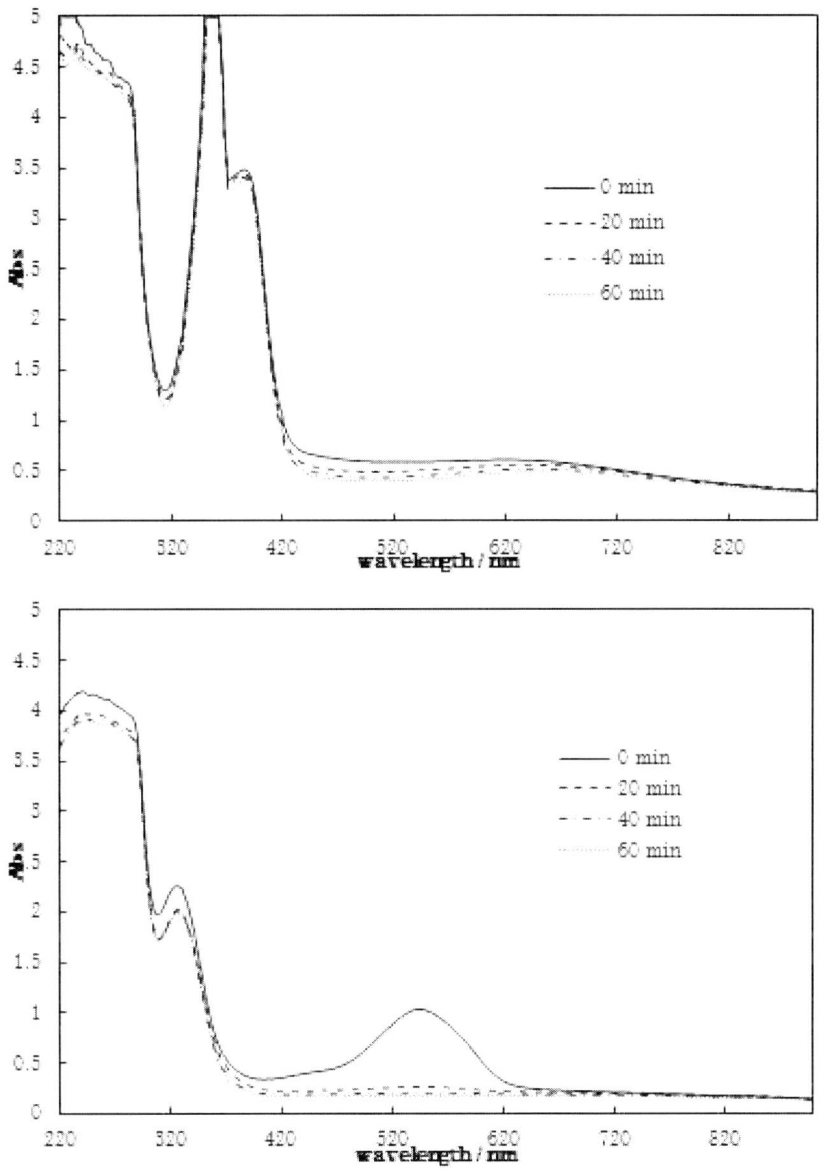

Figure 1. [Up] Changes of UV-vis spectra of a hybrid system of *CuEBSa*, TiO$_2$, and Cr(VI) after reaction by irradiation of visible light for 0, 20, 40, and 60 nm. [Down] UV-vis spectra with a diphenylcarbazide method.

Table 1 summarized the $\Delta\alpha$ rate values. The hybrid systems containing *CuEBSa*, *CuEBDC*, and *CuEBtB*, which could not be reduced by TiO$_2$ and UV light irradiation, can reduce Cr(VI) species up to low concentration in the smaller time than the hybrid system without copper(II) complexes (only TiO_2).

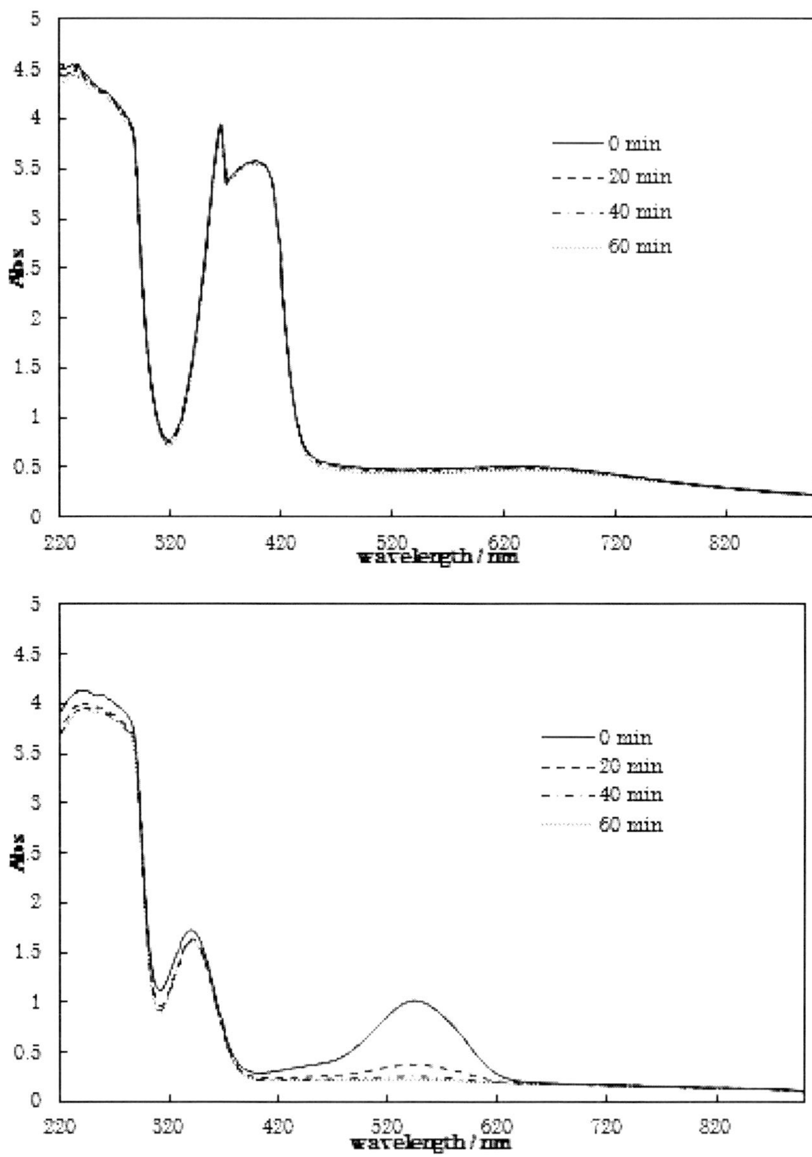

Figure 2. [Up] Changes of UV-vis spectra of a hybrid system of *CuEBDC*, TiO_2, and Cr(VI) after reaction by irradiation of visible light for 0, 20, 40, and 60 nm. [Down] UV-vis spectra with a diphenylcarbazide method.

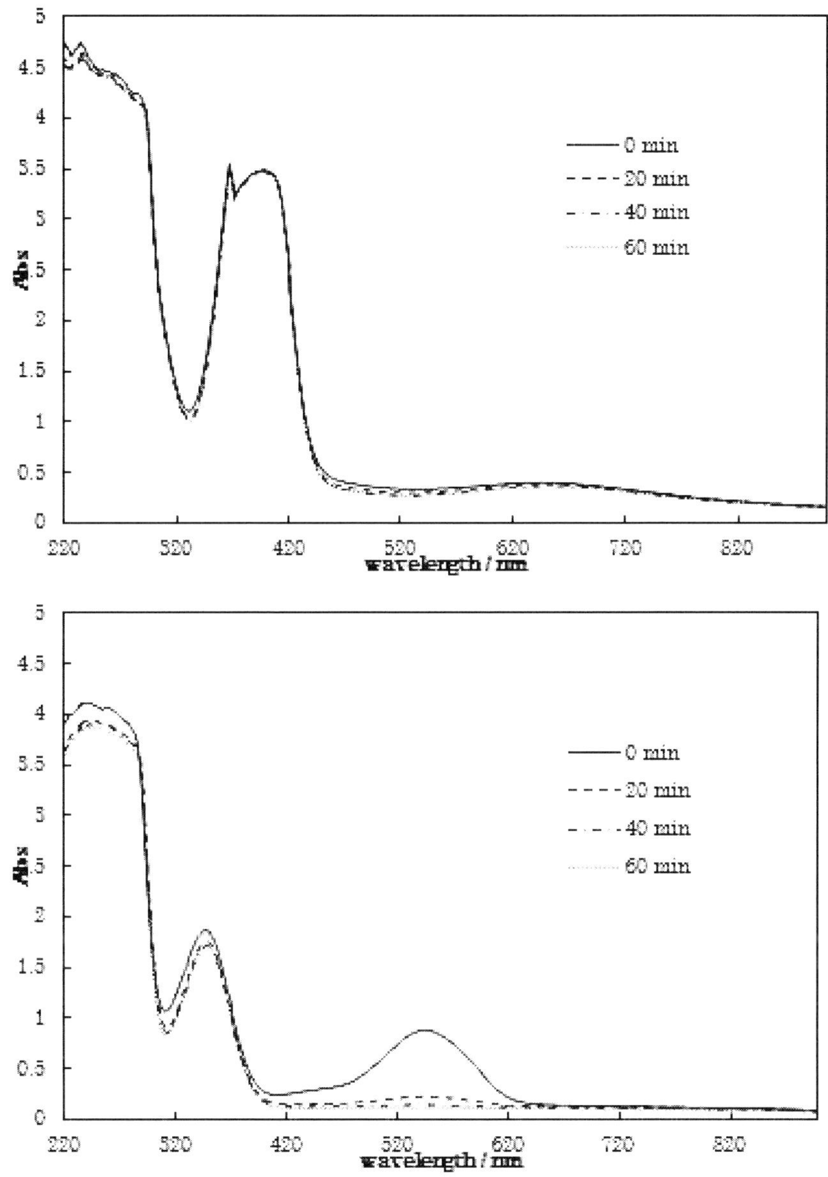

Figure 3. [Up] Changes of UV-vis spectra of a hybrid system of *CuEBtB*, TiO$_2$, and Cr(VI) after reaction by irradiation of visible light for 0, 20, 40, and 60 nm. [Down] UV-vis spectra with a diphenylcarbazide method.

On the other hand, *CuLSa* and *CuLDC*, which *could* be reduced by TiO$_2$ and *UV light* irradiation, could exhibited similar tendency to *CuEBSa,*

CuEBDC, and *CuEBtB*. However, *CuLDC* was not effective, the result of it was close to that of only TiO_2.

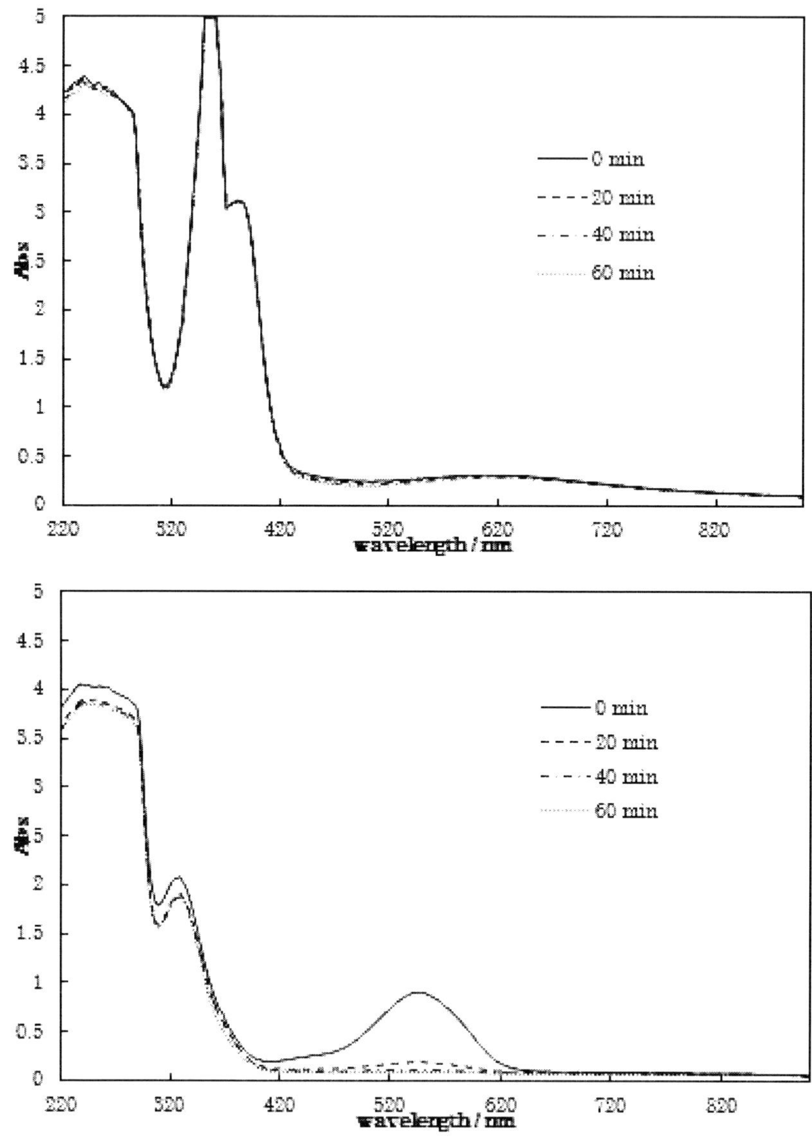

Figure 4. [Up] Changes of UV-vis spectra of a hybrid system of *CuLSa*, TiO_2, and Cr(VI) after reaction by irradiation of visible light for 0, 20, 40, and 60 nm. [Down] UV-vis spectra with a diphenylcarbazide method.

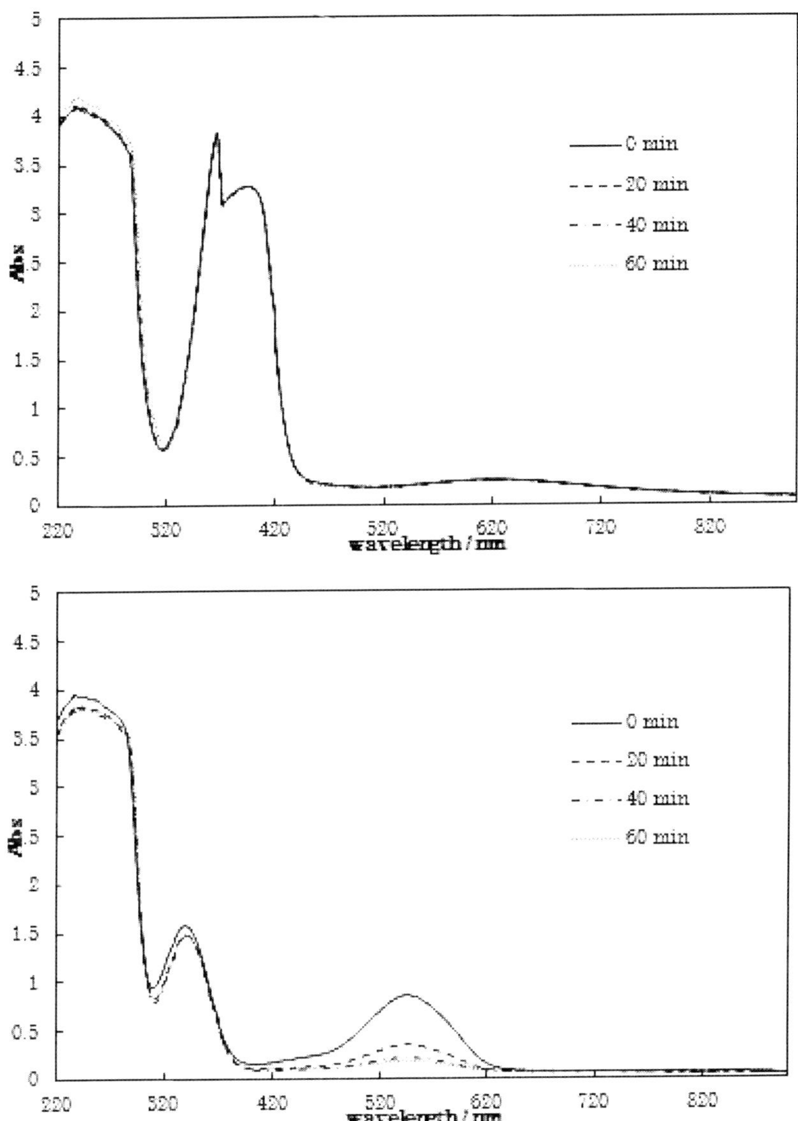

Figure 5. [Up] Changes of UV-vis spectra of a hybrid system of *CuLDC*, TiO_2, and Cr(VI) after reaction by irradiation of visible light for 0, 20, 40, and 60 nm. [Down] UV-vis spectra with a diphenylcarbazide method.

In this way, electrochemical tuning by substituent groups of aldehyde moiety significantly depends on some chemical structures of ligand, *e.g.*, L-amino acid precursors.

Table 1. Change of absorbance at 540 nm (Cr(VI) ions)

	UV light irradiation time			
Complexes	0 min	20min	40 min	60 min
CuEBSa	1	0.109	0.037	0.010
CuEBDC	1	0.210	0.057	0.021
CuEBtB	1	0.155	0.052	0.019
CuLSa	1	0.150	0.053	0.025
CuLDC	1	0.350	0.181	0.136
Non	1	0.370	0.230	0.184

(The leftmost column is labeled $\triangle \alpha$ rate, spanning the complex data rows.)

Non means a control system without copper(II) complex.

In addition, irradiation of *UV light* instead of visible light could occur effective reduction of Cr(VI) species without copper(II) complexes (only TiO_2). Visible light could not occur photo-induced reduction of Cr(VI) species without copper(II) complexes (only TiO_2). Therefore, the following mechanism can be proposed. Since electron excitation of conductive band of TiO_2 can occur by not visible light but UV light, electron transfer from TiO_2 to Cr(VI) could not occur the system without copper(II) complexes (only TiO_2). In contrast to them, the hybrid systems added copper(II) complexes, visible light-excited electron transfer from copper(II) complexes to conductive band of TiO_2 could be suggested and further electron transfer to Cr(VI) species to be reduced as the experimental results exhibited.

Photoreduction of Sm(III) instead of Cr(VI) by hybrid systems. On the other hand, Figure 6 shows UV-vis and fluorescence spectra of a hybrid system containing Sm(III) instead of Cr(VI), *CuLSa*, and TiO_2 before and after visible light irradiation for 40 min.

The reason for selection of *CuLSa* is that it is the most effective reagent for reduction of Cr(VI) ion. Both UV-vis and fluorescence spectra did not exhibit spectral changes before and after visible light irradiation. The result suggested that Sm(III) ion could not be reduced and maintain trivalent species in the present hybrid system.

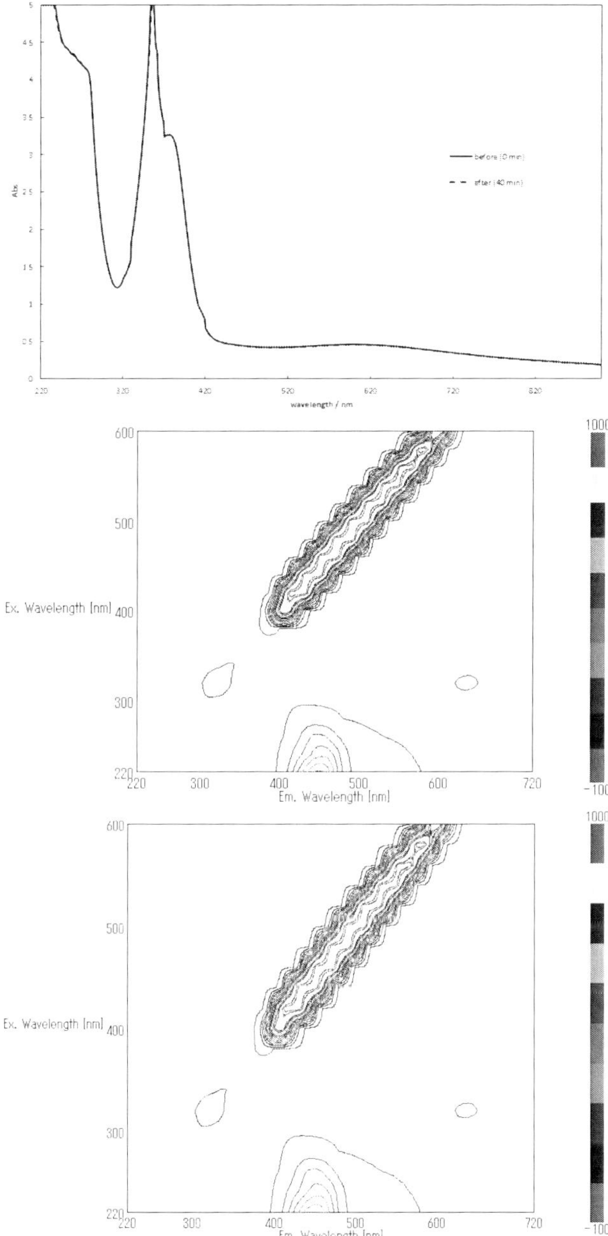

Figure 6. [Up] UV-vis and [down] fluorescence spectra of the hybrid system containing Sm(III) instead of Cr(VI) before and after visible light irradiation for 40 min.

Which attributed that Sm(III) is more difficult to be reduced than Cr(VI) and could not occur electron transfer between TiO_2 under the same experimental condition even if the most effective *CuLSa*.

Examination of working hypothesis of a concept. In the previous work [1], we have proposed "*working hypothesis of a concept*" for the course work. Model hybrid systems composed of TiO_2 microcrystals and chiral copper(II) complexes have been designed and tested for light absorbing effects from the following viewpoints. The results of the present work is added each items. Some new aspects could be found by this work.

1. Chemical reactions just after mixing to obtain hybrid systems.
   No reaction without light irradiation could be observed.
2. Characteristic photo-absorption bands indicating intermolecular interactions (confirming by induced CD bands of chiral transfer or other phenomenon).
   Some copper(II) complexes having well-tuned ligands can react photo-induced electron transfer even by *visible light* irradiation in the hybrid systems.
3. Chemical reactions after irradiation of UV light.
   Of course UV light is more effective for photo-induced electron transfer of TiO_2 directly, visible light can also occur (probably different) chemical reaction for some copper(II) complexes.
4. Matrix environments besides solutions (dispersed into polymers or biomolecules), if it is possible.
   Although environment besides solutions could not be examined even in the present work, the present work became the rare example of reduction of Cr(VI) species in organic solvents (methanol).

CONCLUSION

In summary, we have successfully reduced Cr(VI) ion to Cr(III) up to low concentration after relatively short time *visible light* irradiation by means of hybrid systems of Cr(IV) ion, TiO_2 and copper(II) complexes (*CuEBSa, CuEBDC, CuEBtB*, and *CuLSa*) in *methanol solutions* for the first time. On the other hand, Sm(III) ion could not be reduced and maintain trivalent species in the present hybrid system. Therefore, the present hybrid systems, which have reaction path including visible light excitation and electron transfer concerning appropriate copper(II) complexes, may be effective only for specific metal ions being easy to be reduced.

ACKNOWLEDGMENT

The authors are grateful to The Cosmetology Research Foundation for financial support. They also thank Ms. Haruna Furukawa for her help of Cr(VI) analysis.

REFERENCES

[1]   Yoshida, N.; Akitsu, T. "Reaction of Hybrid Systems Composed of Cu(II) Complexes Having Chiral Schiff Base Amino-Acid Ester Derivatives and TiO_2", Integrating Approach to Photofunctional Hybrid Materials for Energy and the Environment, chapter 5, pp. 111, Nova Science Publishers, Inc., New York, 2013. and references therein.

[2]   Kurata, M.; Yoshida, N.; Fukunaga, S.; Akitsu, T.; *Contemporary Engineering Sciences,* 2013, 6, 255.

[3]   Akitsu, T.; Ishiguro, Y.; Yamamoto, S.; Nishizuru, H. *Asian Chem. Lett.,* 2010, 14, 63.

[4]   Akitsu, T.; Nishizuru, H. *Asian Chem. Lett.,* 2010, 14, 261.

[5]   Watanabe, Y.; Nishizuru, H.; Akitsu, T. *Asian Chem. Lett.,* 2011, 15, 211.

[6]   Nakayama, T.; Minemoto, M.; Nishizuru, H.; Akitsu, T. *Asian Chem. Lett.,* 2011, 15, 215.

[7]   Akitsu, T.; Yamamoto, S. *Asian Chem. Lett.,* 2010, 14, 29.

[8]   Akitsu, T.; Yamamoto, S. *Asian Chem. Lett.,* 2010, 14, 255.

[9]   Yamamoto, S.; Akitsu, T. *Asian Chem. Lett.,* 2011, 15, 203.

[10]   Kyung, H.; Lee, F.; Choi, W. *Environ. Sci. Technol.,* 2005, 39, 2376.

[11]   Sun, B.; Reddy, E. P.; Smirniotis, P. G. *Environ. Sci. Technol.,* 2005, 39, 6251.

In: Samarium   ISBN: 978-1-63321-045-5
Editor: Kaitlyn R. Danford   © 2014 Nova Science Publishers, Inc.

Chapter 5

POTENTIAL APPLICATIONS OF SAMARIUM AS A DOPANT ELEMENT

Nader Shehata[1,2] and Kathleen Meehan[3]*
[1]Department of Engineering Mathematics and Physics,
Faculty of Engineering, Alexandria University, Egypt
[2]The Bradley Department of Electrical and Computer Engineering,
Virginia Tech, US
[3]School of Engineering, University of Glasgow, Scotland, UK

ABSTRACT

Samarium has been studied extensively as a dopant in a wide number of host materials because of its chemical properties and the considerable modification on the properties of the host that are measured when even small concentrations of samarium are present. Samarium has the lowest activation energy of all of the rare earth elements. As a result, the ease at which charged vacancies in an oxide host material such as cerium oxide (ceria) are generated is observed as an increase in the ionic mobility and conductivity of the oxide host. In addition, an increase in the fluorescence intensity has been measured in samarium-doped rare earth oxides when compared to the undoped oxides. The absorbance dispersion of samarium-doped materials exhibits a blue shift into the ultra violet as well. Following a description of the characteristics of samarium-doped hosts, which will include discussion the reasons for the chemical, optical, and structural properties that result from the presence of samarium, a

---

* Corresponding author: Email: nader.shehata@alexu.edu.eg.

review of the applications of the samarium -doped host materials will be presented. These include environmental sensing via electrochemical and fluorescence quenching techniques as well as their potential use as phosphors in solid state lighting and as a nanopharmaceutical diagnostic or therapeutic agent.

1. Review of Samarium Dopant in Inorganic and Organic Hosts

a. Inorganic Hosts

In this part, we are focusing on different types of inorganic hosts where samarium has been used as a dopant. Some ceramic glasses, such as cerium oxide (ceria) and titanium oxide (titania), are considered famous hosts for samarium doping. Some types of inorganic materials including ceramic glasses and large bandgap dielectric fluorite are common hosts for samarium dopant. Titania is a well known as a low-cost and efficient photocatalyst for the detoxication of air and water pollutants with low toxicity up to certain limit of nano-scale size [1-3]. Generally, lanthanide ions with 4f electron configurations when doped in titania crystal may be able to reduce the rate of electron-hole recombination. The lanthanide dopants can lead to shift the wavelength response towards the visible spectrum [4-7].

Doping with samarium ions, as one of the lanthanide ions, has been significantly found to improve the overall photocatalytic activity under UV/visible light irradiation for some degradations such as for methylene blue (MB). Doping titania with Sm can enhance the absorption in the spectrum range between 400 nm and 500 nm. Also, Sm dopant shows a promising resistance to the phase transformation of titania, which can be considered one of the major drawbacks of titania. Related to the photoluminescence (PL) characteristics, it has been recently observed that the stronger the PL intensity, samarium dopant with certain molar ratios can help in increasing the content of oxygen vacancies defects with lower rate of electron-hole recombination. That leads to higher photocatalytic activity of TiO_2 [8].

Related to another famous ceramic host for Sm-dopant, cerium oxide (ceria) can be considered a low-phonon environment for a lot of dopants [9]. Ceria has fluorite-structured ceramic material with no crystallographic phase change at any temperature up to its melting point. Within trivalent dopants, oxygen vacancies are created and increased in the CeO_2 crystal. This property leads to a wide variety of applications, such as oxygen-ion conducting electrolytes in solid oxide fuel cells and oxygen sensors [10]. From the optical characteristics perspective, samarium dopant showed recently promising blue shift in absorbance dispersion and direct bandgap [11]. Also, samarium dopant increases the visible PL emission of ceria under near UV excitation as shown in Figure 2.

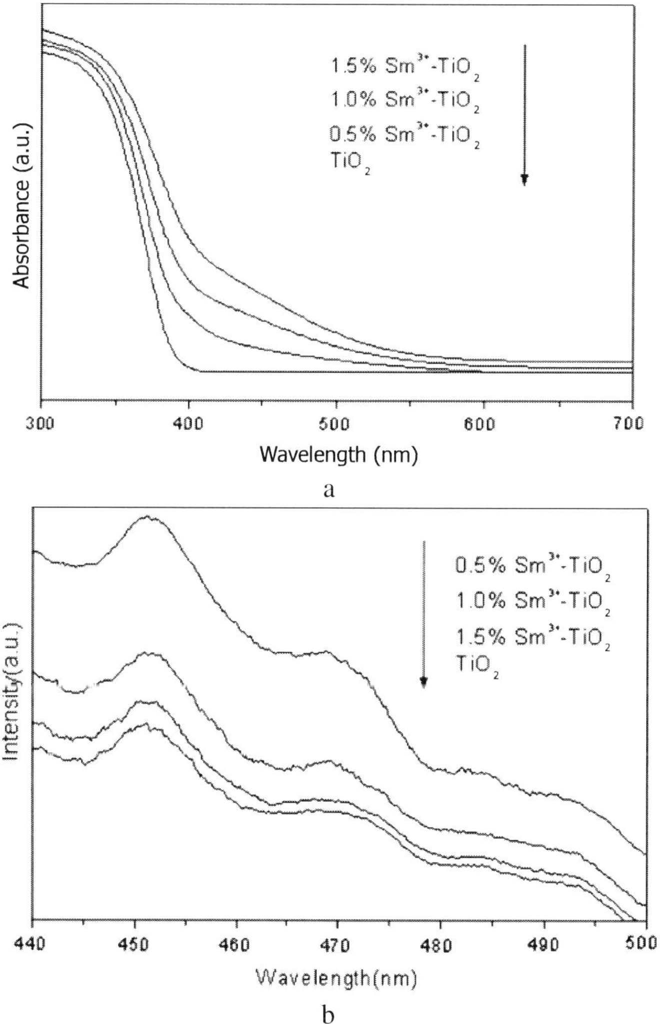

Figure 1. Impact of samarium dopant in TiO_2 through a) optical absorbance and b) PL spectrum at excitation wavelength of 300nm [8].

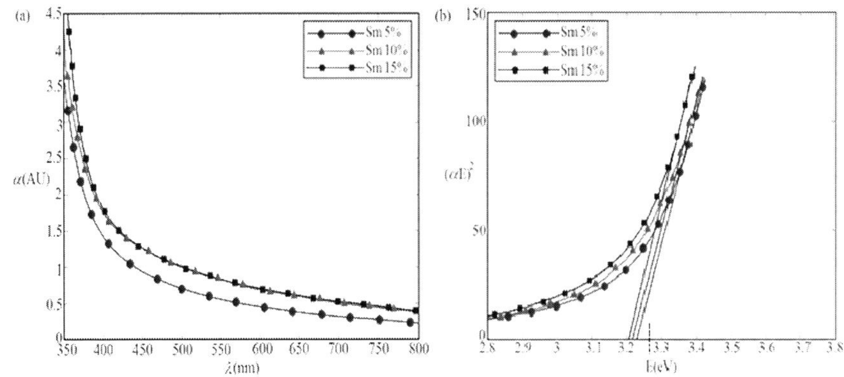

Figure 2. Impact of samarium dopant in CeO_2 through a) optical absorbance and b) direct bandgap [11, 12].

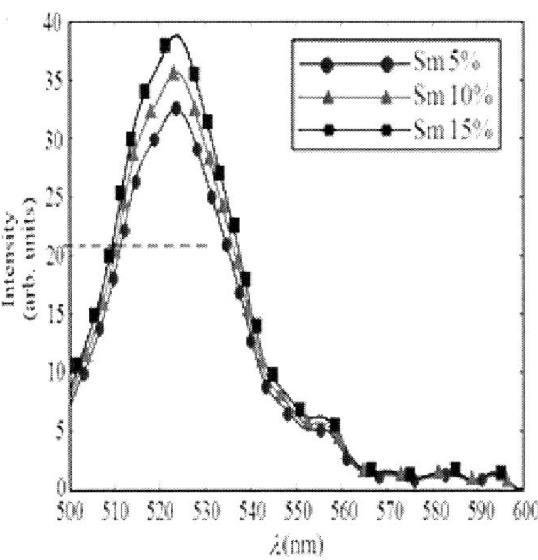

Figure 3. Visible PL spectrum at excitation wavelength of 430nm. The dotted line refers to the fluorescence peak of undoped ceria [12].

Moving to another host category, samarium has been used as a promising dopant in very high bandgap semiconductor. Calcium fluoride (CaF), as an example, is used as a host material with bandgap around 12 eV for samarium dopant. Samarium dopant ions in CaF offers improved number of excited states inside the large forbidden bandgap of the host material. Consequently, the resonant tunneling processes can be found between the formed states to increase the emission current [13]. That can be helpful in the applications related to the one-atom electron source based on sharp conductor tip covered by very large bandgap semiconductor [14], doped with ions as impurity defects in addition the oxygen vacancies formed by such dopants.

b. Organic Hosts

Samarium has been recently used as a dopant in polymers such as polyaniline to improve different characteristics of the organic host. Related to the electrical conductivity, it has been found that the electrical conductivity of the formed composite increases with the increase of Sm-dopant concentration [15]. The increased dopant concentration results in easier movement of charge carriers through the increased localized states. The movement of carriers can be carried out through hopping mechanism. In photoluminescence, samarium dopant has been proved to increase the fluorescent emission at violet colored emissions near to 400 nm. That may be explained due to the interchain species, which is critical in the fluorescence emission of conjugated polymers. The increased concentration of Sm-dopant leads to reduced emitted peak intensities with same PL spectra shape. That might be observed due to the possibility of aggregation of samarium ions in the polymer chain [16]. That means that the peak intensities of the fluorescent emissions can be controlled by embedding samarium dopant at different concentrations.

2. Properties of Sm-Doped Host Materials

a. Physical and Chemical Properties of Sm

Some lanthanide oxides hosts, such as ceria, are doped with trivalent ions to improve its ionic conductivity. As every two trivalent ions that replace tetravalent cerium ions, and consequently an oxygen vacancy is formed to maintain charge neutrality. The resulting lattice has intrinsic charged vacancies defects which are responsible for the increase of ionic conductivity of the host. Indeed, extrinsic defects are formed when a tri-valent dopant is substituted for a Ce^{+4} ion and the tri-valent oxide is instantaneously converted into the tri-valent ion and O-vacancy in the material with the oxygen atoms released from the crystal matrix. This process is shown in the following chemical reaction [17]

$$T_2O_3 \rightarrow 2T'_{Ce} + V_{\ddot{O}} + 3O_O^x$$

where T is the tri-valent element dopant atom, T'_{Ce} means electron-acceptor tri-valent ion formed in ceria, $V_{\ddot{o}}$ is doubly positively charged O-vacancy and O^x_O is the neutral oxygen atom. However, Skala et al. reported that adding tri-valent atoms to ceria not only increases concentration of O-vacancies itself, but also increases the conversion rate from Ce^{+4} to Ce^{+3}[18]. This process leads to a higher concentration of O-vacancies.

Some theoretical studies have recently illustrated the site location impact of samarium dopant in ceria host. These studies use specialized genetic algorithm which shows that low-energy configurations lead to a variety of Sm–vacancy distances from vacancies with at least one Sm ion to be the nearest neighbor (NN) to each vacancy. The recent studies show that samarium doping ions do not exclusively present a particular coordination site around the vacancy which contradicts the traditional next nearest neighbor (NNN) preference obtained from DFT analysis [19].

b. Optical Properties of Sm as a Dopant Ion in Ceramics and Semiconductors

Samarium has been used as a dopant in different optical host materials to be applicable for both optical up- and down-conversions. From the definitions, up-conversion means the emission of higher photon energy such as visible light by excitation source of lower photon energy such as IR. While the optical down-conversion or fluorescence means the emission of lower photon energy from an optical material excited with higher photon energy. Overall, both down- and up-conversions are useful in wide variety of applications such as bio-imaging and gas sensors [20].

In down-conversion, when the host material is excited by relatively large photon energy, then the carriers can be excited to higher energy levels. However, in case of hosts where have defects, there is a large probability of carriers relaxation causing the conduction band electrons to make a transition to the defect state within the bandgap. From the defect state, the electron undergoes multiple transitions in order to return to the ground state according to Shockley Reed Hall recombination. These emissions have lower photon energy than the excitation photons. In this process, samarium plays an important role as a supporter to defects whether by increasing the defects concentration or reducing the activation energy with the vacancies [19]. Therefore, samarium dopant can be helpful in improving the fluorescence peak intensity. Recent research work shows the increase of fluorescent peak intensity with increasing the samarium dopant concentration within some host materials such as ceria nanoparticles as explained previously in Figure 3.

In up-conversion, samarium dopant can be helpful within different hosts used mainly in generating visible or near UV emission from IR or visible excitations, respectively. As described in Figure 4, carriers have to make transitions between certain energy levels in samarium atoms. It has been observed that violet/blue emissions are extracted from Sm-doped some lanthanide oxide nanoparticles such as gadolinium oxide at different excitation optical signals with range between 510 to 710 nm. The clear observed up-converted emission shows a shift with variable excitation source [21, 22].

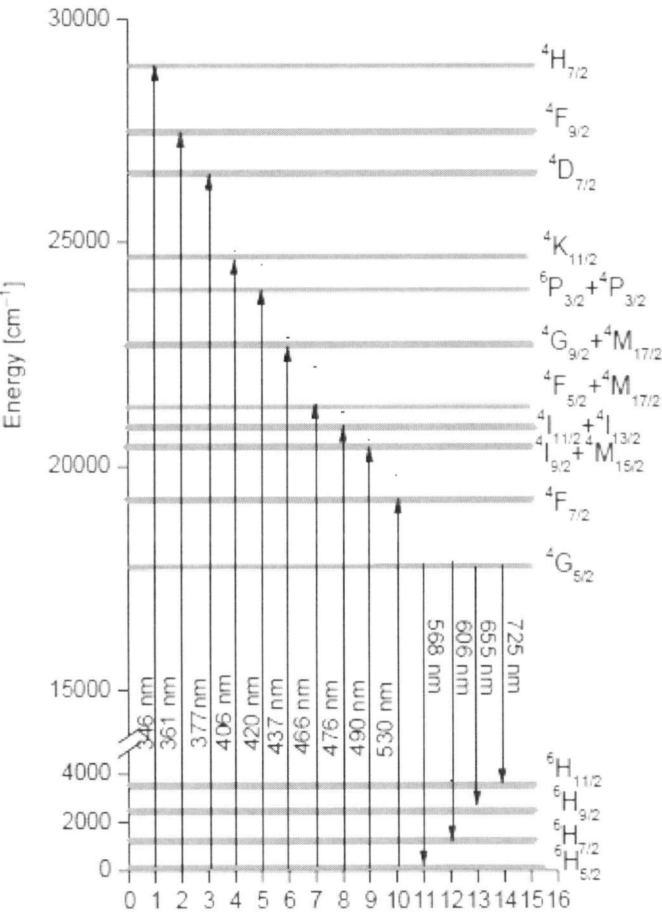

Figure 4. Schematic diagram of Sm transitions for optical up-conversion process [21].

Also, some studies have been conducted on the co-doping of samarium with other lanthanides such as erbium within ceria nanoparticles host. In this case, the alignment of Sm-dopant energy levels with other co-dopant such as erbium can help in up-conversion from IR excitations to visible emissions. In this case, the up-converted emissions have no observed wavelength shift but with controlled peak intensity with varying the Sm-dopant concentration [12].

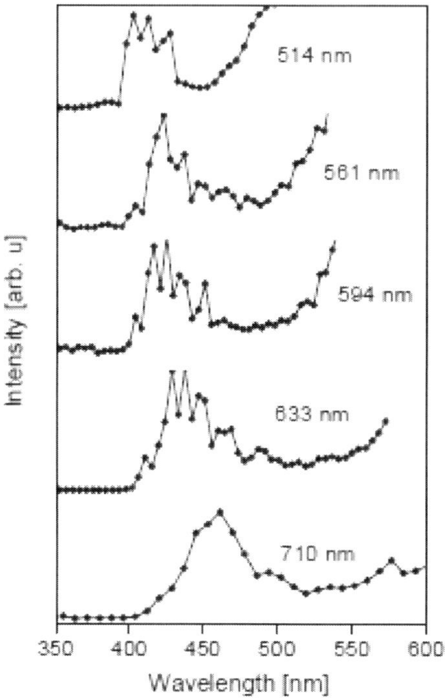

Figure 5. Up-converted emissions of Sm-doped gadolinium oxide [21].

c. Effects of Sintering

There are different techniques for preparing samarium doped hosts for bulk and different nanostructures. For Sm-doped host nanoparticles, the used synthesis methods are chemical precipitation, hydro-thermal synthesis and solid-state reaction method. Compared to most methods, precipitation is attractive due to its use of inexpensive salt precursors, the simple process performed at room temperature and pressure, and the ease at which the process can be scaled for mass production [23-26]. Depending on the sintering technique and conditions, the optical and electrical characteristics can be varied. Also, the structural parameters such as grain size, lattice parameter, and distance between atomic planes are highly sensitive to the synthesis technique [11]. The sintering environment can affect the thermal properties of Sm-doped host which can be detected through some thermal analysis techniques such as thermal gravimetric analyzer (TGA) [27].

d. Electrical Conductivity and Ionic Mobility of Sm-Doped Oxides

The diffusion of charges through Sm-doped materials occurs via a vacancy hopping mechanism [28]. Doping some host types such as ceria with tri-valent atoms usually increases the ionic conductivity compared to undoped ceria because most trivalent ions results in a lower energy barrier for oxygen migration [29]. That leads to higher mobility of oxygen vacancies as explained in the following equation.

$$\sigma = \frac{\sigma_o}{T} \exp(-\frac{E_a}{K_B T})$$

where the ionic conductivity (σ) can be expressed as an exponential function of the activation energy for oxygen vacancy diffusion (E_a). T stands for temperature, K_B for the Boltzman constant and σ_o for a temperature-independent conductivity prefactor. At low dopant content in the ceria system, most of the oxygen vacancies are free. While at high dopant content, the conductivity becomes lower because of the formation of defect associations (e.g., trivalent ion-oxygen vacancy complexes) due to Columbic attraction [17]. Activation energy (E_a) consists of migration energy (E_m), and association energy (E_{ass}), which prevents the O-vacancies of being mobile due to the binding energy between the dopant and the vacancy. E_{ass} becomes the dominant component of E_a within low and intermediate temperatures [19].

It has been shown that Sm^{3+} is the single best dopant to improve the ionic conductivity of different oxide. In details, samarium has been proved to increase the concentration of the O-vacancies in doped lanathanide oxides, with having relatively low activation energy with the charged vacancies. That helps in increasing the ionic conductivity through the hopping mechanism over the low-activated vacancies.

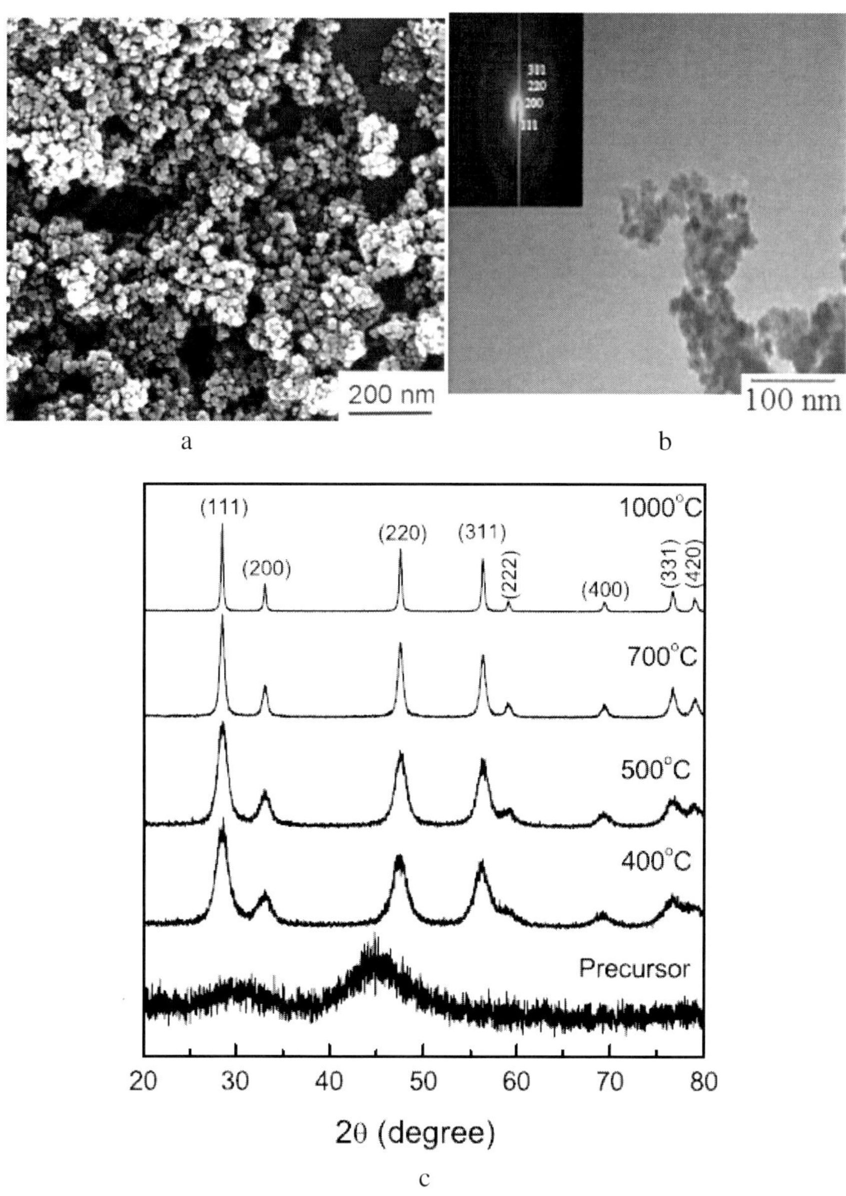

Figure 6. a) TEM image of Sm-doped ceria nanoparticles prepared at room temperature for sintering time up to 24 hours and b) FESEM image of Sm-doped ceria powder morphology calcinated at 700°C for 2 hours. and c) XRD pattern of Sm-doped ceria powder at different calcinations temperature [11, 27].

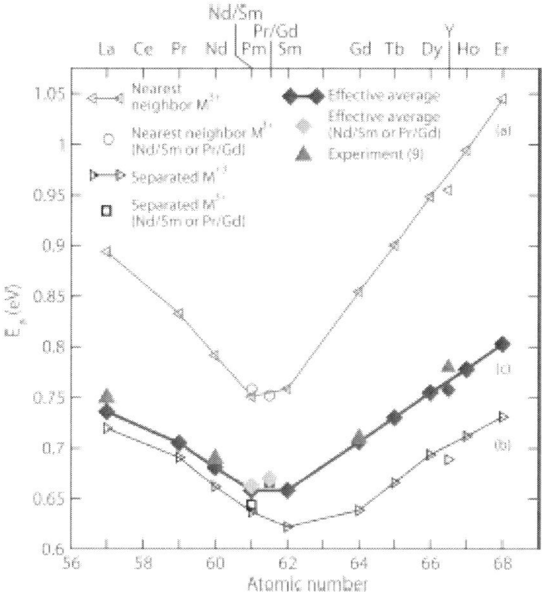

Figure 7. Activation energy (Ea) for different lanthanide dopants with vacancies. This Figure shows that samarium has lowest Ea between the other lanthanides [19].

e. Magnetic and Dielectric Properties of Sm-Doped Hosts

Samarium has two ionic states: Sm^{2+}, where the doubly-ionized samarium has an even number of 4f electrons, and Sm^{3+} with an odd number of electrons in the 4f orbital. Thus, Sm^{3+} is magnetic and Sm^{2+} is not. A further complication is that Sm^{3+} has a number of low-energy excited multiplet states. As a result, the ground state of Sm^{3+} may not be equal to J= 5/2, which alters the magnetic susceptibility of the compound. In addition, stress applied to samarium compounds can cause change in the ionic charge of samarium, converting Sm^{2+} to Sm^{3+} with an accompanying change in magnetism of the material [30]. Tuning of the dielectric properties as well as the magnetic properties in this manner is possible [31]. Hence, the magnetic properties of Sm-based compounds can be quite complex. $SmPd_2Al_3$ is a good example. Its magnetic properties are highly anisotropic as well as temperature dependent, exhibiting four meta-magnetic states as an applied magnetic field increases from $0 - 0.75$ T at a temperature of 1.8 K [32].

The properties of Sm-doped materials are also of interest. $BiFeO_3$ is a multiferroic material, one in which more than one ferroic order (antiferromagnetic, ferromagnetic, and ferroelectric) can exist at the same time in a single phase. The magnetism in such materials can be tuned by changing the magnitude of an applied electric field and, similarly, the electrical properties of the material can be altered by applying a magnetic field. When Sm replaced some of the Bi atoms in this material, the resulting crystalline structure is a more perfect crystal with improved ferromagnetic properties and lower dielectric loss compared to undoped $BiFeO_3$ [33]. Magnetic hysteresis was observed at room temperature in this material as well as in $Bi_7Fe_3Ti_3O_{21}$ [34], where Sm was substituted for some of the Bi, indicating that a magnetically ordered state is present at this temperature. The density and grain size of multicrystalline Sm-doped materials has been found to increase compared to that of the undoped materials in certain cases [35] and decrease in others [36]. As one might expect some of these trends correlate with the difference in ionic radius between the dopant, Sm, and the atom that it replaces, distorting the local crystal lattice. However, the charge state of the Sm complicates the data analysis.

3. APPLICATIONS OF SAMARIUM-DOPED CERAMIC MATERIALS

a. Electrochemical and Optical Environmental Sensing

The development of oxygen sensing has had profound effects in the fields of medical science, bioengineering, environmental monitoring, and industrial process control, and in military applications [37]. Research to develop smaller and more accurate sensors that are robust in harsh or reactive environments continues. In biomedicine as an example, the development of a biomedically-compatible microscale dissolved oxygen (DO) sensing system, using nanotechnology methods, has been extensively pursued in recent research studies for applications that include in vivo and in vitro oxygen sensing especially prior to and during surgical procedures, such as during tumor removal operations [38]; probing the surfaces of transplant tissues and organs [39]; and monitoring blood gas levels in microdialysis units [40].

The bulk of the sensor technology falls under three classifications: fluorescence-dye optical sensing, amperometric electrochemical sensing, and thin film solid-state conductivity sensing. The electrical sensing technique for oxygen is the most reliable and the wider used in commercial sensors.

Resistive-type metal–oxide semiconducting oxygen sensors operate on the principle of change in their characteristic resistance by the interaction of the sensing surface with the oxygen gas under ambient conditions [41-42]. Sensitivity in these platforms can be enhanced by doping the metal oxides with some lanthanides, which improves both the carrier concentration and mobility. Recent research work shows the conductivity change of Sm-doped host materials (such as ceria thin-films) with changing the oxygen pressure, temperature and film thickness. The oxygen pressure or temperature can affect on the conductance of the Sm-doped ceria (SDC) thin-film. The measured current is a function of oxygen pressure and temperature at film thickness range between 50-300 nm [15].

Although the resistive sensors are the most reliable, this type of sensors has some limitations such as fouling by organic matter, which leads to catastrophic failure and short operational lifetimes [63]. Additionally, electrochemical sensors require a reference electrode, which had been proved to be difficult to miniaturize and the characteristic electrostatic charge of this electrode attracts charged particles and ions, causing the device to be extremely prone to contamination by organic matter [64]. So, the optical sensing technique, which is less reliable, but can overcome some of the disadvantages of the electric technique.

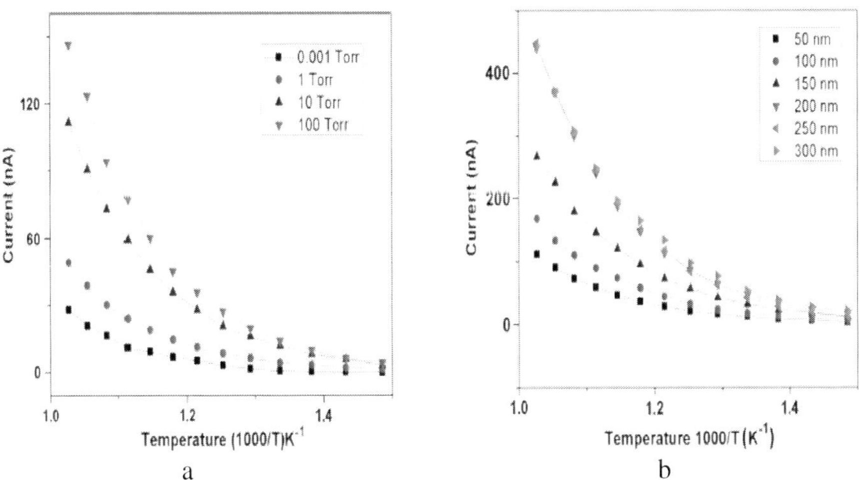

Figure 8. Variation of total measured current with changing the temperature (a) at different oxygen pressure (at constant thin film thickness of SDC of 50nm), (b) at variable film thickness (at constant oxygen pressure of 10 torr) [15].

The main optical phenomenon employed when sensing dissolved oxygen using fluorescent dyes is fluorescence quenching, caused by an energy transfer to the dissolved oxygen followed by the formation of non-radiative complexes, when oxygen interacts with the dye molecule. Fluorescence-quenching oxygen sensors have been adopted rapidly due to their inherent high sensitivity and stability [43].

A majority of fluorescent dyes, particularly the ones that are organometallic in nature, displays oxygen-quenching characteristics. However, a discrete set of fluorophores exhibits enhanced sensitivity to oxygen, as well as relatively long fluorescent lifetimes, thus reducing the requirements of the optical transducers [44]. Examples include some of the ruthenium-based dyes [45-46]; however, several of these dyes are sensitive to water, which negatively impacts the accuracy of the measurement of dissolved oxygen in aqueous solutions.

To overcome this issue, polymers doped with fluorescent dyes have been formulated and these composites have been used as the active material in some fluorescence-based dissolved oxygen (DO) sensors [47]. Recent research work shows Sm-doped optical host such as ceria nanoparticles can be used as optical DO sensor. The main advantage is the improved oxygen vacancies which can act as oxygen probe in addition to the enhanced fluorescence emission [20].

Fluorescence emission decreases with increasing DO concentration in the fluorescent dye where the amplitude of the emitted fluorescence intensity is related to the DO concentration through Stern-Volmer equation, which is as follows [48]:

$$\frac{I_o}{I} = 1 + K_{SV}[O_2]$$

where I_o and I represent the steady-state fluorescence intensities in the absence and presence, respectively, of the quencher; oxygen; K_{SV} is the Stern–Volmer quenching constant, which is an indication for the sensitivity of the used nanoparticles to sense the dissolved oxygen [49]; and $[O_2]$ is the dissolved oxygen concentration. Samarium is shown to be a promising dopant in ceria nanoparticles to increase the visible emission peak intensity and its Stern-Volmer constant up to 370 M^{-1}, compared to undoped ceria nanoparticles and other nanomaterials. Also, the sensitivity to DO is proved to be highly stable with temperature variations up to 50°C [20].

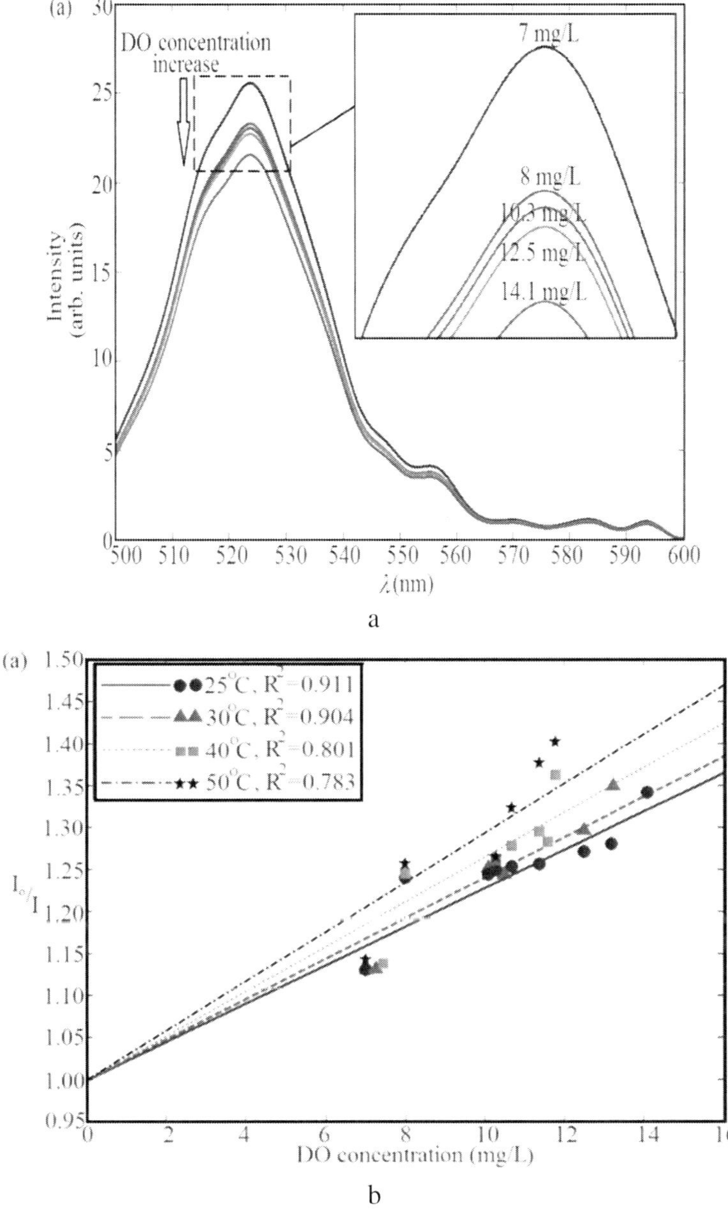

Figure 9. a) The variations of fluorescence peak intensity of Sm-doped ceria nanoparticles at different DO concentrations and b) Stern-Volmer relations with temperature change [20].

b. Solid Oxide Fuel Cells

The production of current as a result of the oxidation of reactants at a temperature well above room temperature (400 $^{\circ}$C or higher) is the basis of solid oxide fuel cells (SOFCs) [50]. The current that is produced from the electrochemical reaction can be composed of electrons, protons, and/or ions, which flows between the anode and cathode of the SOFC through an electrolyte – the solid oxide – that lies between the two electrodes. The solid oxide also may support the catalysis of the oxidation process along oxide-gas interfaces. The electrolyte is typically a rare-earth oxide or ceramic, which is composed of materials such as yttria-stablized zironia (YSZ) or ceria [51]. The high operating temperature of the SOFC is dictated by the material properties of the electrolyte as well as the reaction kinetics. The electrical conductivity of the solid oxides decreases with increasing temperature, which improves the efficiency of the SOFC. Unfortunately, there are a number of disadvantages to such systems; a primary one is that the systems must be connected to a second source of energy. The start-up time for a SOFC can be 20 minutes or longer as the second source of energy is used to bring the SOFC up to its operating temperature. The temperature of the system must be slowly lowered to prevent crack formation in the electrolyte or other components as the SOFC is shut down. Poisoning of the electrolyte can more easily occur at raised temperatures so steps must be taken to insure the purity of the reactants with active or passive removal of impurities such as sulfur -based contaminants. As a result, the current SOFCs are not viable as an alternative energy generator in residential or mobile applications.

The material properties of the electrolyte must be improved to reduce the temperature of operation of SOFCs without sacrificing the efficiency and reliability of the fuel cell [52]. One critical issue is the resistance of the electrolyte, which must be low at temperatures closer to room temperature without compromising the robustness of the solid oxide. This can be accomplished by changing the physical geometry of the electrolyte, reducing its thickness and/or increasing its cross-sectional area. However, these changes may reduce the lifetime of the electrolyte due to increased mechanical stresses due to thermal expansion of the oxide and thermal mismatch between the electrolyte and the electrodes. Alternatively, techniques to modify the material properties of the electrolyte itself have a greater potential to reduce the operating temperature of SOFCs. These techniques include the increase in the ionic conductivity of the existing ceramics through the intentional introduction of impurities in the oxides via doping during synthesis [53]; the improved control over the grain size of the crystallites within oxide [54-57], the interfaces between these crystallite [58, 59], and the interfaces between the oxide and the electrodes[53, 60]; and the development of new electrolytic materials3. Dope ceria is a leading candidate as a solid-oxide electrolye for intermediate-temperature and low-temperature SOFCs.

The charge transport in ceria is determined by the combined transport of oxygen vacancies, protons, and electrons in the material, where the mobility and concentration of each charge carrier is affected by the temperature of the material, the presence of impurities within each grain of the material as well as impurities along the boundaries between grains, and the exchange of charge at these interfaces among other factors. A number of studies have shown that doping ceria can alter the ionic conductivity of ceria by modifying the concentration of oxygen vacancies, as discussed previously. A number of rare earth ions can significantly increase the concentration of oxygen vacancies [11], leading to an increase in conductivity of the doped ceria when compared to that of undoped ceria. Doping ceria with Sm causes the largest increase in ionic conductivity of all of the rare-earth elements because of an increase in stability of the fluoride structure of CeO_2 with a high concentration of oxygen vacancies [61]. Further increases in the conductivity can be obtained by co-doped ceria with a second impurity such as yttrium, which has been shown to increase the mobility of oxygen vacancies [62].

A disadvantage of Sm-doped ceria is that the cycling of ionic states between Ce^{3+} and Ce^{4+} is critical element of the catalytic activity of ceria. Thus, doping ceria with Sm (or Y) has been shown to suppress its catalytic activity for certain reactions [63, 64]. An approach to recover or increase the catalytic activity of Sm-doped ceria is to form electrolyte from a mixture of metal and doped ceria nanoparticles, where the metal nanoparticles also act as catalysts [65].

c. Nanopharmaceutical Diagnostic and Therapeutic Agents

The use of rare earth elements in biomedical applications has a long history. Given their magnetic and optical properties along with the heavy atomic weights and the availability of radioisotopes, both organic and inorganic materials containing one or more rare earths have been used as contrast agents in a number of imaging modalities including MRI, x-ray, and up-conversion luminescence imaging [66], as fluorescent and/or magnetic tags bound to biomolecules and cells, optically- or magnetically-triggered thermal agents used to eradicate cancer in hyperthermia treatments, and as radiotherapeutic agents [67]. Other biomedical applications leverage the effect that these rare earth materials have on biochemical processes and cells, which can potentially lead to their use as antibacterial agents [68] and anticoagulants [69, 70].

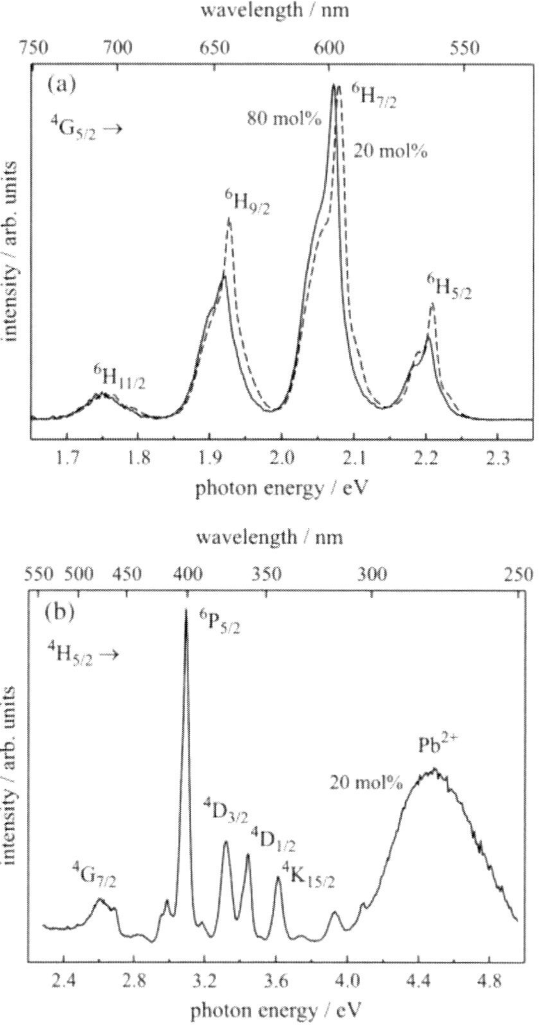

Figure 10. (a) PL emission and (b) PL excitation of Sm^{3+}-doped lead borate glasses; the excitation was carried out at 401 nm; the emission was detected at 598 nm. The emission spectra are normalized to maximum intensity [86].

Samarium-based materials have been studied for use in a number of biomedical applications. In dentistry, samarium compounds have been used as a filter in oral x-ray imaging with the goal to reduce the dosage received by patients during bitewing radiographs [71]. Samarium absorbs both low energy and high energy x-ray photons that do not contribute to imaging process with a 40% reduction in the dosage to which patients were exposed [72]. Samarium-cobalt magnets have been evaluated as a means to apply magnetic forces that can hold dentures, bridges, and plates in place [72] as well as to realign teeth [73]. Studies have shown that there can be corrosion of the magnets with use and potential cell toxicity from dissociated metals, although sealing the metals appeared to be sufficient to eliminate the concern [74, 75].

^{153}Sm is a neutral radioisotope that emitted beta-particles with a half-life of approximately 47 hours [77]. In complexes of samarium-153-ethylenediaminetetramethylenephosphonic acid [(153)Sm-EDTMP] or samarium-153-lexidronam (Quadramet®), ^{153}Sm preferentially accumulates at bone surfaces [78, 79]. These compounds have been shown to reduce the pain associated with metastic bone cancer [79-81] and rheumatoid arthritis [82] and have been approved by the US FDA for this use. Studies have begun to evaluate other therapeutic applications for this radioisotope including its use to treat osteosarcoma [83] and in brachytherpy [84]. More recent study has evaluated samarium oxide as a possible material to replace bone [85].

Sm^{3+} fluoresces in the visible wavelength range when excited by near-UV or UV photons [86], as shown in Figure10 (a). Unfortunately, the excitation wavelength (Figure 10 (b) [86]) is strongly absorbed by biological tissue, which limits the *in-vivo* applications of samarium-labeled fluorescent probes. Sm^{3+}-doped polystyrene and silica nanoparticles have been used as molecular probes for multi-target immunoassays [87, 88]. The detection limits obtained for Sm^{3+}-labeled probes were about six times higher than for Eu^{3+} [81], likely due to the higher quantum efficiency of the radiative process in Eu^{3+} as compared to Sm^{3+} [89].

Interest in improved imaging techniques of living tissue remains high and in the ability to track bimolecules and cells *in-vivo* remains high. Up conversion luminescence (UCL) imaging relies on a two-photon (or more) absorption process to obtain emission from several rare earths, overcoming the issues with absorption of the near-UV or UV excitation source by proteins and water. UCL has been used to obtain images from living tissues from deep within the organism. The efficiency of up conversion in Sm^{3+} is host-dependent; however, red emission has been observed when excited with infrared light with much weak peaks seen in the green and orange regions of the visible spectrum (Figure 13) [90]. However, other rare earth ions have high up-conversion efficiencies [61]. Some researchers are investigating co-doping of nanoparticles with Sm^{3+} and another rare earth such as Eu^{3+}, Yb^{3+} or Tm^{3+} for multimodal imaging, where the properties of the other rare earth are used when imaging via UCL and the radioactivity of ^{153}Sm is employed in single-photon emission computer tomography (SPECT) [91].

While a number of the metal oxide particles have been found to be toxic when the particle size is less than roughly 100 nm [92, 93], ceria nanoparticles have exhibited an ability to extend cell longevity, reducing oxidative stress by scavenging free radicals, in many types of cell lines [94, 95]. It is an open question whether the physciochemical behavior of ceria *in-vivo* will be altered by the presence of Sm.

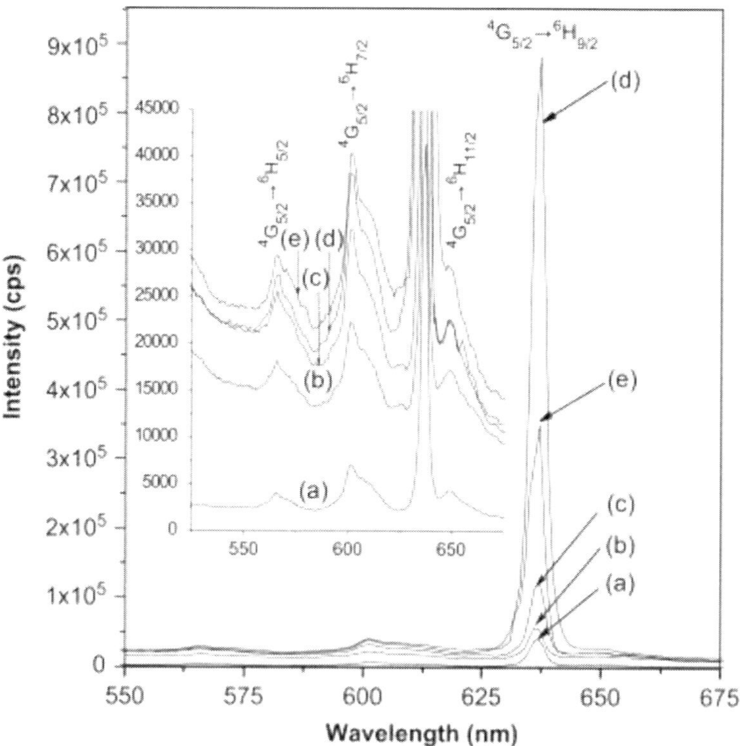

Figure 11. Upconversion fluorescence spectra of: (a) 0.1, (b) 0.3, (c) 0.5, (d) 0.7 and (e) 1.0 wt% Sm_2O_3-doped $15K_2O–15B_2O_3–70Sb_2O_3$ (mol%) glasses upon excitation at λ_{ex}=949 nm radiation [90].

REFERENCES

[1] Hoffmann, M. R., Choi, S. T., Martin, W. & Bahnemann, D. W. (1995). *Chem Rev*, *95*, 69.

[2] Fujishima, A., Rao, T. N. & Truk, D. A. (2000). *J Photochem Photobiol C: Photochem Rev*, *1*, 1.

[3] Choi, W., Termin, A. & Hoffmann, M. R. (1994). *J Phys Chem*, *98*, 13669.

[4] Li, F. B., Li, X. Z., Hou, M. F., Cheah, K. W. & Choy, W. C. H. (2005). *Appl. Catal A: Gen*, *285*, 181.

[5] Zhang, Y., Xu, H., Xu, Y., Zhang, H. & Wang, Y. (2005). *J Photochem. Photobiol A: Chem*, *170*, 279.

[6] Yan, X., He, J., Evans, D. G., Duan, X. & Zhu, Y. (2005). *Appl Catal B: Environ*, *55*, 243.

[7] Xie, Y., Yuan, C., Li, X. (2005). *Mater Sci Eng B*, *117*, 325.

[8] Xiao, Q., Si, Z., Zhang, J., Xiao, C., Yu, Z. & Qiu, G. (2007). *J. Mater. Sci.*, *42*, 9194.

[9] Guo, H. J. (2007). *Solid State Chem*, *180*, 127.

[10] Li, J. G., Ikegami, T. & Mori, T. (2004). *Acta Materialia*, *52*, 2221.

[11] Shehata, N., Meehan, K., Hudait, M. & Jain, N. J. (2012). *Nanopart Res.*, *14*, 1173.

[12] Shehata, N. (2012). PhD dissertation, submitted to Virginia Tech, November.

[13] Gadzuk, J. W. (1993). *Phys. Rev. B*, *47*, 12832.

[14] Konopsky, V. N., Sekatskii, K. S. & Letokhov, V. S. (1996). *J. Physique IV Colloque C5, suppldment au Journal de Physique III*, *6*, 125.

[15] Sanghavi, R., Nandasiri, M., Kuchibhatla, S., Jiang, W., Varga, T., Nachimuthu, P., Engelhard, M. H., Shutthanandan, V., Thevuthasan, S., Kayani, A. & Prasad, S. (2011). *IEEE Sensors J.*, *11*, 217.

[16] Ameen, S., Ali, V., Zulfequar, M., Haq, M. M. & Husain, M. (2009). *J. Appl. Polymer Sci.*, *112*, 2315.

[17] Cioatera, N., Parvulescu, V., Rolle, A. & Vannier, R. N. (2010). Int. Semicond. Conf., 317.

[18] Skala, T., Tsud, N., Prince, K. C. & Matolin, V. (2011). *App. Surf. Sci.*, *257*, 3682.

[19] Andersson, D. A., Simak, S. I., Skorodumova, N. V., Abrikosov, A. & Johansson, B. (2006). *PNAS*, *83*, 1 (Copyright (2006) National Academy of Sciences, U.S.A).

[20] Shehata, N., Meehan, K., Ashry, I., Kandas, I. & Xu Y. (2013). *Sens Actuators B, 183*, 179.

[21] Goldys, E. M., Drozdowicz-Tomsia, K., Zhu, G., Yu, H., Jinjun, S., Motlan, M. & Godlewski, M. (2006). *Optica Applicata, 36.*

[22] Treadaway, M. J. & Powell, R. C. (1975). *Physical Review B: Solid State, 11*, 862.

[23] Suda, E., Pacaud, B. & Mori, M. (2006). *J. Alloys Compd., 408*, 1161–1164.

[24] Dikmen, S., Shuk, P. & Greenblatt, M. (1999). *Solid State Ion., 126*, 89.

[25] Yamamura, H., Takeda, S. & Kakinuma, K. (2007). *Solid State Ion., 178*, 1059.

[26] Dhannia, T., Jayalekshmi, S., Kumar, M. C. S., Rao, T. P. & Bose, A. C. (2009). *J. Phys. Chem. Solids, 70*, 1443.

[27] Ji-Guang Li, Takayasu Ikegami & Toshiyuki Mori (2004). *Acta Materialia, 52*, 2221.

[28] Nolan, M., Fearon, J. E. & Watson, G. W. (2006). *Solid State Ion., 177*, 3069.

[29] Matovic, B., Dohcevic-Mitrovic, Z., Radovic, M., Brankovic, Z., Brankovic, G., Boskovic, S. & Popovic, Z. V. (2009*). J. Power Sources, 193*, 146.

[30] "The Unusual Magnetic Properties of Samarium and its Compounds", The European Synchrotron, http://www.esrf.eu/UsersAndScience/ Publications/Highlights/2005/HRRS/HRRS4 [viewed online: March 16, 2014]

[31] Li, D. & Subramanian, M. A. (2000). *Solid State Sci., 2*, 507.

[32] Pospíšil, J., Kratochvílová, M., Prokleška, J., Diviš, M. & Sechovský, V. (2010). *Phys. Rev. B, 81*, 024413.

[33] Ain, R. S. N., Hakim, S. A. & Hashin, M. (2012). *Adv. Mater. Res., 501*, 329.

[34] Bućko, M., Polnar, J., Lis, J., Przewoźnik, J., Gąska, K. & Kapusta, C. (2013). *Adv. Sci. Technol., 77*, 220.

[35] Wang, F., Wang, X., Zhu, J. & Xiang, L. (2011). *Mater. Sci. Forum., 675-677*, 143.

[36] Jo, S. K., Park, J. S. & Han, Y. H. (2010). *J. Alloys Compounds, 501*, 259.

[37] Wolfbeis, O. S. (1991). *Fiber Optic Chemical Sensors and Biosensors*, CRC Press, Boca Raton, 13-17.

[38] Masson, N. & Ratcliffe, P. J. (2004). *Meth. Enzym., 381*, 305.

[39] Dittmar, A., Mangin, S., Ruban, C., Newman, W. H., Bowman, H. F., Dupuis, V., Delhomme, G., Shram, N. F., Cespuglio, R., Jaffrezic-Renault, N., Roussel, P., Barbier, D. & Martelet, C. (1997). *Sens. Act. B*, *44*, 316.

[40] Cooney, C. G., Towe, B. C. & Eyster, C. R. (2000). *Sens. Act. B*, *69*, 183.

[41] Xu, Y. & Zhou, X. (2000). *Sens. Act. B: Chem.*, *65*, 2.

[42] Ivers-Tiffée, E; et al., (2001). *Electrochimica Acta*, *47*, 807.

[43] Chang-Yen, D. A. & Gale, B. K. (2003). *Lab Chip*, *3*, 297.

[44] Crosby, G. A. & Watts, R. J. (1971). *J. Am. Chem. Soc.*, *93*, 3184.

[45] Lin, C. T., Bottcher, W., Chou, M., Creutz, C. & Sutin, N. (1976). *J. Am. Chem. Soc.*, *98*, 6536.

[46] Hartmann, P., Ziegler, W., Holst, G. & Lubbers, D. W. (1997). *Sens. Act. B*, *38*, 110.

[47] Douglas, P. & Eaton, K. (2002). *Sens. Act. B*, *82*, 200.

[48] Chu, C. S. & Lo, Y. L. (2008). *IEEE Photo. Tech. Lett.*, *20*, 63.

[49] Iosin, M., Canpean, V. & Astilean, S. (2011). *J. Photochem. Photobio. A*, *217*, 395.

[50] Park, S., Vohs, J. M. & Gorte, R. J. (2000). *Nature*, *404*, 265.

[51] Steele, B. C. H. & Heinzel, A. (2001). *Nature*, *414*, 345.

[52] Hibino, A. Hashimoto, T. Inoue, J.-I. Tokuno, Yoshida, S.-I. & Sano, M. (2000). *Science*, *288*, 2031.

[53] Andersson, D. A., Simak, S. I., Skorodumova, N. V., Abrikosova, I. A. & Johansson, B. (2006). *PNAS*, *103*, 3518.

[54] Kim, S. & Maier, J. (2002). J. *Electrochem. Soc.*, *149*, J73.

[55] Kim, S. & Maier, J. (2004). *J. Eur. Ceram. Soc.*, *24*, 1919.

[56] Tschope, A., Sommer, E. & Birringer, R. (2001). *Solid State Ionics*, *139*, 255.

[57] Tschöpe, A. Kilassonia, S. & Birringer, R. (2004). *Solid State Ionics*, *173*, 57.

[58] Li, Z.-P., Mori, T., Aucheterlonie, G. J., Zou, J., Drennan, J. & Miyayama, M. (2011). *Solid State Ionics*, *191*, 55.

[59] Yan, P., Mineshige, A., Mori, T., Wu, Y., Auchterlonie, G. J., Zou, J. & Drennan, J. (2013). *ACS Appl. Mater. Interf.*, *5*, 5307.

[60] Knibbe, R., Hjelm, J., Menon, M., Pryds, N., Søgaard, M., Wang, H. J. & Neufeld, K. (2010). *J. Am. Cer. Soc. 93*, 2877.

[61] Eguchi, K., Setoguchi, T., Inoue, T. & Arai, H. (1992). *Solid State Ionics*, *52*, 165.

[62] Venkatesh, V. & Reddy, C. V. (2013). *J. Mod. Phys.*, *4*, 1499.

[63] Zhao, S. & Gorte, R. J. (2004). *Appl. Catalysis A: Gen.*, *277*, 129.

[64] Farra, R., García-Melchor, M., Eichelbaum, M., Hashagen, M., Frandsen, W., Allan, J., Girgsdies, F., Szentmiklósi, L., López, N. &. Teschner, D. (2013). *ACS Catalysis*, *3*, 2256.

[65] Saeki, M. J., Uchida, H. & Watanabe, M. (1994). *Catalytic Lett.*, *26*, 149.doi.org/10. 1007/BF00824040

[66] Gai, S., Li, C., Yang, P. & Lin, J. (2014). *Chem. Rev.*, *114*, 2343. doi: 10. 1021/cr4001594

[67] de Patricio, B. F. C., de Albernaz, M. S., Sarcinelli, M. A., de Carvalho, S. M., Santos-Oliveira, R. & Weissmüller, G. (2014). *J. Biomed. Nanotechnol.*, *10*, 1242.

[68] Wang, Y. J., Hu, R. D., Jiang, D. H., Zhang, P. H., Lin, Q. Y. & Wang, Y. Y. (2011). *J. Fluoresc.*, *21*, 813.

[69] Funakoshi, T., Furushima, K., Shimada, H. & Kojima, S. (1992). *Biochem. Int.*, *28*, 113.

[70] Ying, H. U., Lei, Z. & Yi, D. (2013). *Acta Metallurgica Sinica*, *6*, 756.

[71] Mauriello, S. M., Matteson, S. R., Tyndall, D. A. and Bader, J. D. (1989). Oral Surg. *Oral Med. Oral Pathol.*, *68*, 108.

[72] Gelskey, D. E. & Baker, C. G. (1981). *Oral Surg. Oral Med. Oral Pathol.*, *52*, 565.

[73] Hirata, M. (1977). *Kokubyo Gakkai Zasshi*, *64*, 544.

[74] Bondemark, L. (1994). *Swed. Dent. J. Suppl.*, *99*, 1.

[75] Bondemark, L., Kurol, J. & Wennberg, A. (1994). *Br. J. Orthod.*, *21*, 335.

[76] Boeckler, A. F., Ehring, C., Morton, D., Geis-Gerstorfer, J. & Setz, J. M. (2009). *J. Prosthodent.*, *18*, 301.

[77] Lee, M. R. & Katz, R. (1954). *Physical Review*, *93*, 155.

[78] Ogawa, K. & Washiyama, K. (2012). *Curr. Med. Chem.*, *19*, 3290.

[79] Lam, M. G., de Klerk, J. M., van Rijk, P. P. & Zonnenberg, B. A. (2007). *Anti-cancer Agents in Med. Chem.*, *7*, 381.

[80] Sartor, O. (2004). *Rev. Urol*, *6*, S3.

[81] Liepe, K., Runge, R. & Kotzerke, J. (2005). *J. Cancer Res. Clin. Oncol.*, *131*, 60.

[82] Siegel, H. J., Luck, J. V. Jr. & Siegel, M. E. (2004). *J. Am. Acad. Orthop. Surg.*, *12*, 55.

[83] Senthamizhchelvan, S., Song, H., Frey, E. C., Zhang, Z., Armour, E., Wahl, R. L., Loeb, D. M., & Sgouros, G. (2012). *J. Nucl. Med.*, *53*, 215.

[84] Valente, E. S. & Campos, T. P. (2010). *Appl. Radiat. Isot.*, *68*, 2157.

[85] Herath, H. M. T. U., Di Silvio, L. & Evans, J. R. G. (2010). *J. Biomed. Mater. Res.*, *94A*, 130.

[86] Dyrba, M., Krause, S., Phau, C., Micle, P.-T. & Schweizer, S. (2013). *Radiation Measurements*, *56*, 36.

[87] Huhtinen, P., Kivelä, M., Kuronen, O., Hagren, V., Takalo, H., Tenhu, H., Lövgren, T. & Härmä, H. (2005). *Anal. Chem.*, *77*, 2643.

[88] Murray, K., Cao, Y. C., Ali, S. & Hanley, Q. (2010). Analyst, 135, 2132.

[89] Bador, R., Déchaud, H., Claustrat, F. & Desuzinges, C. (1987). *Clin. Chem.*, *33*, 48.

[90] Som, T. & Karamakar, B. (2008). J. *Lumines.*, *128*, 1989.

[91] Wu, Y., Sun, Y., Zhu, X., Liu, Q., Cao, T., Peng, J., Yang, Y., Feng, W. & Li, F. (2014). *Biomater.*, *35*, 4699.

[92] I. L. Bergin & F. A. Witzmann, (2013). *Int. J. Biomed. Nansci. Nanotechnol.*, *3*, 163

[93] K. Liu, X. Lin, & J. Zhao, (2013). *Int. J. Nanomedicine*, *8*, 2509.

[94] B. A. Rzigalinski, (2005). *Technol. Cancer Res. Treat.*, *4*, 651.

[95] B. A. Rzigalinski, K. Meehan, R. M. Davis, Y. Xu, W. C. Miles, & C. A. Cohen, (2006). *Nanomedicine*, *1*, 399.

INDEX

E

F